Strategies for Sustainability

Series Editors

Rodrigo Lozano, Faculty of Engineering and Sustainable Development, University of Gävle, Gävle, Gävleborgs Län, Sweden

Angela Carpenter, Faculty of Engineering and Environment, University of Gävle, Gävle, Sweden

Subseries: Spatial Planning and Sustainable Development

Subseries Editor

Zhenjiang Shen, School of Environmental Design, Kanazawa University 2C718, Kanazawa City, Japan

Strategies for Sustainability

The series focuses on implementation strategies and responses to environmental problems at the local, national, and global levels.

Our objective is to encourage policy proposals and prescriptive thinking on topics such as: sustainability management, sustainability strategies, lifestyle changes, regional approaches, organisational changes for sustainability, educational approaches, pollution prevention, clean technologies, multilateral treaty-making, sustainability guidelines and standards, sustainability assessment and reporting, the role of scientific analysis in decision-making, implementation of public-private partnerships for resource management, regulatory enforcement, and approaches to meeting inter-generational obligations regarding the management of common resources..

We favour trans-disciplinary perspectives and analyses grounded in careful, comparative studies of practice, demonstrations, or policy reforms. This largely excludes further documentation of problems, and prescriptive pieces that are not grounded in practice, or sustainability studies. Philosophically, we prefer an open-minded pragmatism – "show us what works and why" – rather than a bias toward a theory of the liberal state (i.e. "command-and-control") or a theory of markets. We invite contributions that are innovative, creative, and go beyond the 'business as usual' approaches.

We invite Authors to submit manuscripts that:

- Prescribe how to do better at incorporating concerns about sustainability into public policy and private action.
- Document what has and has not worked in practice.
- Describe what should be tried next to promote greater sustainability in natural resource management, energy production, housing design and development, industrial reorganization, infrastructure planning, land use, and business strategy, and organisational changes.
- Develop implementation strategies and examine the effectiveness of specific sustainability strategies.
- Focus on trans-disciplinary analyses grounded in careful, comparative studies of practice or policy reform.
- Provide an approach "…to meeting the needs of the present without compromising the ability of future generations to meet their own needs," and do this in a way that balances the goal of economic development with due consideration for environmental protection, social progress, and individual rights..

Themes covered in the series are:

Sustainability management
Sustainability strategies
Lifestyle changes
Regional approaches
Organisational changes for sustainability
Educational approaches
Pollution prevention
Clean technologies
Multilateral treaty-making
Sustainability guidelines and standards
Sustainability assessment and reporting
The role of scientific analysis in decision-making
Implementation of public-private partnerships for resource management
Governance and regulatory enforcement
Approaches to meeting inter-generational obligations regarding the management of common resources

More information about this series at http://www.springer.com/series/8584

Stephen Siu Yu Lau • Junjie Li
Shimeng Hao • Shuai Lu

Editors

Design and Technological Applications in Sustainable Architecture

The perspective of China, Japan, Singapore and Thailand

 Springer

Editors
Stephen Siu Yu Lau
Shenzhen University
Guangdong, China

Shimeng Hao
Beijing University of Civil Engineering
and Architecture
Beijing, China

Junjie Li
Beijing Jiaotong University
Beijing, China

Shuai Lu
University of Sydney
Darlington, NSW, Australia

ISSN 2212-5450   ISSN 2452-1582   (electronic)
Strategies for Sustainability
ISBN 978-3-030-80033-8   ISBN 978-3-030-80034-5   (eBook)
https://doi.org/10.1007/978-3-030-80034-5

This Springer imprint is published by the registered company Springer Nature Switzerland AG
The registered company address is: Gewerbestrasse 11, 6330 Cham, Switzerland

Preface

The arrival of the Anthropocene era (Crutzen & Stoermer, 2000) has had a significant impact on everyone on Earth. The agenda for Anthropocene as a recognized entry to Earth's history by the scientific (geographic) community recognizes those responsibilities of human activities, their adverse impacts on climate change and ultimately, the equilibrium of every life form and species on this Planet. More recently, numerous research communities have linked the cause and damage of Earth's environment and lifeforms to the building and construction sector. The United Nations and other major international bodies have attributed global responsibility to the building and construction industry at as much as 30 percent, if not higher. In this context, members of the industry have been called upon to examine the specific role of architecture and building construction in order to alleviate environmental depletion. In the recent decade, the theme of sustainable or regenerative design has continued to capture market attention and stirred immense interests and provoked constructive debates and discussions among both the profession and academia as well as other stakeholders. Technology-based architectural design strategies and paradigms, in broad-ranging related topics, for example, climate adaptability and responsiveness, heat island, biodiversity, ecological, and bio-restoration, evolved at fast pace and accumulated a body of knowledge that serves as beacon for practitioners and researchers in building design, construction, usage, operation, and maintenance. The scope of such growing knowledge entails a cross-disciplinary fermentation as it encompasses those interdisciplinary aspects of social, economic, humanistic, and environmental perspectives beyond the usual realm of engineering and design. In essence, it covers a broad scale – ranging from the scale of urban planning to building and finally down to the building detail. In essence, the surge for new knowledge has driven authors to join forces together, to make available frontier research and practicum for knowledge gathering, synthesis, and dissemination. More recently, the Spatial Planning and Sustainable Development Group SPSD based in Kanazawa, Japan, has operated a collaborative platform for the international research community. It has organized a conference in Singapore where researchers have come together for deliberation, sharing, and debating. Majority of the topics in this book have been selected from this event. This book is a forum for the testing of new and old ideas, mostly in the field of architecture and construction in a host of countries – China, Japan, Singapore, and Thailand, which share similar but varied cultural and environmental features. From these geographic boundaries, the book invites scholars and practitioners of architecture, engineering, landscaping, and planning as well as biologist and social scientists interested in the built environment, from all over the world, to come together on the topic of climate responsiveness of sustainable architecture design and technology. The topics are extended from urban planning and building design to performance evaluation.

From the Editorial Team: LAU, LI, LV, and HAO

Guangdong, China   Stephen Siu Yu Lau
Beijing, China   Junjie Li
Beijing, China   Shimeng Hao
Darlington, NSW, Australia   Shuai Lu

Contents

Part III   Health and Human Considerations

Part IV   Building Performance and Design Evaluation

Chapter 1
Introduction – On the Definition and Measure of Sustainability

Stephen Siu Yu Lau

It all started from an emergent concept known to most researchers as "the pillars of sustainability" which appeared on the discussion table in and around the late 1990s. A normalized way to apprehend the objective of such a concept is achieved through three realms of considerations, that is, economics, society and environment. For some, the three pillars are better known as "the triple bottom line" which represents the baseline approach in order to measure the status of sustainable development. Both definitions draw reference to the ultimate goal of sustainable development, or sustainability. For those of us who are in pursuit of the idealism of sustainability would not forget the toils and doubts those who called themselves green advocates encountered on their pilgrimage to sustainability! Yes, suspicion that persistently devours both the mind and will of the pursuer appeared out of nowhere but to occupy the mind like a never-ending oxymoron conflict caused by these pillars acting on one another, if not oneself. Among them, it is argued that economics is the first and ethically sinful pillar of all. For instance, how much would it cost extra to change the production plant in the name of environmental sustainability? How many years would it take in order for a new sustainable technology to yield fiscal benefits (speaking of renewable solar energy, for instance)? Queries such as these fill the brain until there is no space left for an answer with good cause. The second pillar, which is the social one, is even harder to explain because it is at its core all about divergences such as idiosyncrasy, preference, culture, habit, and, more so, individual like and dislike. How could one change lifestyles instantaneously? Similarly, how difficult is it to change or modify various habits of the building user, such as convincing office workers to climb four stories in both directions at their workplaces instead of traveling by elevators? The final one is the seemingly smallest and

S. S. Y. Lau (✉)
Department of Architecture, School of Design & Environment, National University of
Singapore, Singapore, Singapore

© The Author(s), under exclusive license to Springer Nature Switzerland AG 2021
S. S. Y. Lau et al. (eds.), *Design and Technological Applications in Sustainable
Architecture*, Strategies for Sustainability, https://doi.org/10.1007/978-3-030-80034-5_1

weakest third pillar, that of environmental consideration. Better known as environmental sustainability, it is deemed an achievable goal by effecting specific interventions to prevailing building design norms and practices, in order to produce not a "negative" but "positive" energy building, for instance. Allegedly, buildings are referred to as negative buildings because their existence impels a ceaseless consumption of multitude forms of resources of which energy is but one (material, water, land, daylight, environmental qualities, and numerous indicators for health are others). Unfortunately, the embrace of mainstream green by academia and practicum is far from satisfactory, partly because of the above struggles. At this juncture, it becomes clear for readers to realize how interwinding and contradictory conflicts of the three pillars have indeed hampered the path to triumph (where it literally meant arriving at the apex of the three pillars that marks the state of equilibrium for sustainable development). Some researchers have distinguished "the pillars of reality" from "the pillars of idealism" by "the fallacy of the three pillars" (Ciegis et al., 2009; Kohon, 2018; Purvis et al., 2019) in order to remind everyone of a collective and an individual responsibility to overcoming such complexes of difficulties and contradictory challenges in the way of sustainable development. As explained by others, Fig. 1.1, the fallacy of the three pillars is the first step in the pursuit of sustainable development, i.e. to realize the state of chaos and struggles because of the three considerations. The second step as it were is represented by those who prefer to stick with the baseline approach – by upholding each of the baselines as corresponding threshold (the bottom line) for measuring economic, social and environmental attainments at the national, institutional or corporate level. Known as TBL or 3BL, it is a kind of accounting system for checking or boosting performance in the respective realm of concerns. For most policymakers or investors, the 3BL approach is synonymous with the concept of "equilibrium," achievable by the "appropriate interactions" (sustainable development) of "planet" (environment), "people" (social) and "profit" (economics). Arguably, the third and important step is proposed by the author and others to be the conscientious revelation, or more

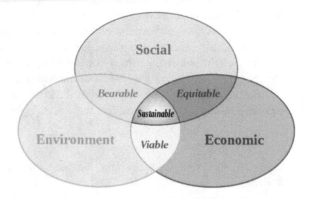

Fig. 1.1 The three pillars of the sustainability concept: social-environmental-economic. (Source: Wikipedia)

commonly known as corporate social responsibility (CSR). The call for an expression of the willingness or undertaking of social responsibilities has come about in relative recent times as above debates and discussions continued. In no time, the debates initiated a revisit of the caring awareness – the need to care for others as a rider over other concerns. In fact, such topic could be dated back to the 1930s by the Berle and Dodd debate. In today's sharing context, much of the discussions could be referred to business ethics, or business sustainability. In essence, it introduces good practice such as "beyond compliance," "philanthropic responsibility" and "self-regulation of the private business enterprises," and "supply chain management."

In an idealized world, could academicians from around the world who are preoccupied by an intellectual pursuit of meanings projected by theory through its naming be found? On the official agenda for debate are the names of "green" and "sustainable." Green or sustainable as names for buildings that embrace sustainability by design, construction, and operation have gone through rounds of dialectic discussions over embedded meanings and projected implications (e.g., vision as well as marketability) of the concept the name represents. Those who are in favor of "green" held the opinion that color green is the most appealing wavelength for the human eyes. In a physiological way it justifies the name because green is the color that brings comfort to the beholder. Others believe in the human instinctive affinity with natural elements (nature-relatedness) and associate such affinity with the color green. As a name, it projects the image of a building for comfort, so why not green as the official name? A common complaint about green buildings (even today) has been the saying that green buildings are ugly buildings. At the time an architect writer Lance Hosey, in a 2007 article, entitled "The Good, the Bad… Where is the design in sustainable design?" summarized some of the gist of both accusation and counteraccusation directly and indirectly associated with the message projected by such naming. He wrote, "the ugly truth about green building is that much of it is ugly." For the author, the justification for such a remark is built upon the critique of a bias on science and technology, and or the lack of emphasis on art and related matters as inspiration or driver for design. Here one is being reminded of the distinction between building and architecture, one which is based on technology and the other on creativity and art, in simplified terms. Hosey reacted to the criticism that the green design paradigm was based on nothing but science and technology, and with it a mandate realized by an evaluative template that hindered the production of creativity otherwise in the design process. At the end of the writing, he responded to the question(s) posed in the beginning of the essay regarding "ugliness" and the criticism of a neglect of aesthetics by quoting journalist David Orr, "if it's not beautiful, it's not sustainable" as a counter argument in defense of green or sustainable building. For the sake of our discussion here, one sees the interchangeable use of green and sustainable by various stakeholders as though the two represent the same contents and concerns of sustainability. Reference to various definitions support the interchangeability of the twin name as a practice, whether green or sustainable. It suffices that green is being used in connection with building design, while sustainable is concerned with resources utilization for building, and its environmental impacts on the immediate and overall environment. In other words, green refers to the building itself while sustainable is concerned with the environmental impacts from buildings onto their neighbors. It is not too late when another attempt on the naming issue surfaced and expanded the scope of the discussion. This time the proposal centered on the legitimacy of "high-performance building" as an alternative name. As it was reasoned, the name bears relevance to the reduced use of resources by enhancing the overall performance of a building as the primary target, hence the terminology high-performance building serves the best descriptor in lieu of green or sustainable building! Other than the emphasis on performance, there also came the newer and allegedly appropriate name such as "Low-carbon building" focusing on carbon as an ultimate measure and achiever of sustainability for every form of human activity including the building sector. While performance directs attention to one specific aspect of expectations of building (that of functionality and efficiency), carbon draws attention towards a scientific interpretation and understanding of environmental consequences and impacts for the building sector, which explains all measurable items by carbon as a singular unit of measure of the so called "good" or "bad." In reality, scientists have been frantically working on translating every output and measurable into units of carbon on a daily basis across the world. Before the day when the translation has come about, another concept has just emerged as a nominee for the naming campaign. As the name implies, "positive energy building" is allegedly a transformative achievement – referring to the colossal technical and administrative efforts necessary for its realization. Namely, the name refers to the transformation of the role of building(s), from a negative consumer of energy to that a positive producer of energy. For instance, positive energy district is a relatively recent buzz word for the concerted production of clean energy at the scale of district or community. Technically, a Positive Energy District (PED) is "based on a high level of energy efficiency, in order to keep annual local energy consumption lower than the amount of locally produced renewable energy" (cityxchange.au, 2020).

A corollary is the naming of Anthropocene, also known as the Human Epoch, as a proposed candidate for registration in the Earth's geological history. As reported, the effort has taken more than a decade (started 2008) in order for a consensus to be reached by respective communities of scientists of the world to finally agree and adopt the naming, going through a systematic protocol. In many ways, it is scientists' share of resolving the ongoing global debate on what or who is ultimately responsible for climate change that results in unprecedented deterioration of the environment, posing severe threats to all forms of life on the planet. At the heart of the debate, whether climate change is nature caused or not demands a scientific inquiry into the cause – how and when it started, while working out the remedies simultaneously at a global scale.

In conclusion, looking back at the trails one has taken and those intense discussions that have taken place, it is not hard to realize that firstly the root cause for some of the confusion or arbitrariness is associated with the many attempts to find the proper or correct representation of the concept being named after, and secondly that those efforts that have gone into the tedious naming debate exercises are in fact necessary as each of the attempts are efforts to seek the truth by a rational and logical way. Similar to the geological scientists' attempt at finding the truth about Earth's history, we, on the other hand, have spent no lesser time in our own debates seeking the same truth – of sustainability. Last but not least, as an introduction to the writings this book contains, it also serves as a reflection of the ways we have lived with the companion of the environment amid many of the good and bad things that come with such co-existence of the two. Some have argued for the adopting of eco-logical building (or city) as the proper name to represent the symbiosis that have happened one way or another through the passage of time for man and nature on this planet. Attaching with the discipline of ecology as a name for what and how we have wanted (designed) our buildings and the built environment to be, calling them ecological buildings or ecological city repeatedly reminds us of the significance of such symbiotic relationship of man as an element within nature and not the other way round. For the authors in this book, such a realization is the crux of the many debates that have gone on, and perhaps will continue. It is important to show that we care, and want to care.

References

Ciegis, R., Ramanauskiene, J., & Martinkus, B. (2009). The concept of sustainable development and its use for sustainability scenarios. *Engineering Economics, 62*(2), 28–37.

CityXChange. (2020). Positive Energy Districts (PED). Retrieved 20 August 2021, from https://cityxchange.eu/knowledge-base/positive-energy-districtsped/

Kohon, J. (2018). Social inclusion in the sustainable neighborhood? Idealism of urban social sus-tainability theory complicated by realities of community planning practice. *City, Culture and Society, 15*, 14–22.

Purvis, B., Mao, Y., & Robinson, D. (2019). Three pillars of sustainability: In search of conceptual origins. *Sustainability Science, 14*(3), 681–695.

Part I
City Ecological Restoration

Chapter 2
Study on Mechanism of Climate Response and Ecological Restoration in High-Density Urban Environment

Dexuan Song and Yi Liu

2.1 What Is the High-Density Urban Environment

For the "high density" in the word "high-density urban environment," it contains two connotations: high population density and high building density. Population density refers to the number of people in a certain unit space, while the building density can refer to the building area or total area in a certain unit space. It should be noted that density is a comparative concept. There is no absolute standard for high or low density. It is closely related to the environment it is located in.

With the increase in population and the enhancement of construction capacity, high density is an inevitable trend of current urban development. However, many scholars believe that high density is a problem that should be resolved, as it will inevitably bring oppression (Aflaki et al., 2017; Radhi et al., 2013; Xue & Luo, 2015; Zhao et al., 2010). This view is not valid. In a well-developed high-density urban area, an "oppressive" environment may not necessarily be formed (Fahy & Cinnéide, 2008; Jenks & Jones, 2009; Yang, 2008). Conversely, this intensive use of land will improve transportation efficiency, reduce urban energy consumption, slow down the endless expansion of the city, and maintain the ecological environment of the city (Fig. 2.1).

The high-density urban environment is composed of the internal environment of the building and the external environment of the building. Among them, the external environment is not a natural environment, but a designed one based on natural conditions and user needs, combined with the building. Compared with the internal environment of the building, the external environment is not only affected by artificial construction but also more directly by the natural environment such as sunlight,

D. Song (✉) · Y. Liu
College of Architecture and Urban Planning, Tongji University, Shanghai, China
e-mail: dxsong@tongji.edu.cn

Fig. 2.1 Aerial view of Shanghai Lujiazui. (Photo by Denys Nevozhai on Unsplash)

wind, and rainfall. In this chapter, we will focus on the external environment in the high-density urban environment. Without special instructions, the high-density urban environment in this chapter refers exclusively to the external environment of the building.

According to the types of constituent elements, the high-density urban environment can be divided into natural ecology subsystems, artificial construction subsystems, and social humanity subsystems. The natural ecology subsystem covers non-biological elements such as water, soil, air, wind, light, sound, radiation, climate, and biological elements such as plants and animals. The element quality of the natural ecology subsystem is an important indicator for evaluating the ecological quality of high-density urban environment. The artificial construction subsystem mainly includes road traffic, hard landscape, municipal equipment, and building facades. Its construction level affects the quality of daily life of citizens to a large extent. The social humanity subsystem covers the non-material elements that have potential or even direct impacts on the urban environment and citizens' lives, such as urban services and management, pension environment, and cultural environment.

2.2 Ecological Restoration

2.2.1 From "Ecological Recovery" to "Ecological Restoration"

There are many explanations for the connotation of "Ecological Restoration," including "Ecological Recovery," "Ecological Reconstruction," "Ecological Renewal," "Ecological Renovate," etc. (Bradshaw & Wong, 2003; Song & Zhou, 2001).

Although using the same word "ecological restoration," those early researches are more limited to "Recovery" rather than "Restoration." It can be concluded that ecological recovery refers to following the rules of ecosystem succession, relying on its own organization and regulation ability or few artificial regulations means to restore the function of the ecosystem.

The concept of ecological restoration was put forward slightly later than that of ecological recovery, but they are often confused. The International Society for Ecological Restoration defines and distinguishes recovery and restoration. SER puts forward the criterion for judging whether an ecosystem is restored is "whether the rich biological and non-biological resources in the ecosystem can continue to develop without human assistance" (SER, 2004). It is further explained that both restoration and recovery use a certain pre-existing ecosystem as a reference, but the goals and strategies adopted by the two are different. The restoration focuses on ecosystem processes, productivity, and ecology system services; however, recovery also emphasizes the reconstruction of the previously existing biota, that is, species composition and community structure.

In this article, "ecological restoration" refers to the concept of ecological restoration in a broad sense. The ecological restoration of the high-density urban environment specifically refers to the restoration of damaged or degraded existing ecological elements through artificial means of modern technology and in accordance with the objective laws of nature, thereby restoring the natural ecosystem, promoting the rational circulation and flow of material and energy, reducing negative environmental interference, and meeting the sustainable development of the environment.

2.2.2 Ecological Restoration in Built Environment

The research on the built environment was relatively early in Europe and the United States. In the 1970s, developed countries had already solved the housing needs of the people, leading needs to transform from material to spiritual. The construction of outdoor environment received primary attention, and people began to be aware of the impact of the external environment on the quality of life.

In "the Warsaw Declaration of Architects" issued at the UIA conference in 1981, it was mentioned that architects and planners should recognize the relationship between human beings, architecture, and the environment, which also promoted architecture, environment, planning, and ecology (International Union of Architects, 1981). Multidisciplinary research has promoted the diversified research on the external environment of buildings.

Experts and scholars pay more attention to the exploration of urban sustainability, citizen participation in urban development, and promotion of urban development by urban institutions. The Soviet ecologist O. Yanitsky put forward the concept of "eco-city" in 1987 (Yanitsky, 1987), emphasizing the establishment of human ecological settlements with efficient use of material, energy, and information, coordinated development of society, economy, and environment, and a virtuous cycle of ecological elements. On the basis of this concept, the United States, Germany, Singapore, Norway, Sweden, and other countries have formulated environmental protection and ecological technology systems and begun to shift to ecological environment industries that focus on natural ecological restoration. The American ecologists Richard Forman and Michel Godron published the book *Landscape ecology*, which studied the basic structural patterns of ecology and began to use the principles of ecology to guide the exploration of planning and design practice (Forman & Godron, 1986). Based on the basic principles of the ecological structural model and the model language "patches-corridor-matrix," they complete model discrimination, structure, and function analysis, which complements the shortcomings of Mcharg's melaleuca-style planning and design model, such as being limited to the vertical ecological process. The book *Landscape Architecture: A Manual of Site Planning and Design* by the American designer J. O. Simonds in 1997 put forward systematic planning principles and design models for specific environmental issues. The main content involves environmental factors and corresponding planning techniques, such as cluster planning, analysis of ecological determinants, urban renewal, urban and regional planning structure, water and air protection, etc., combined with the case of Tres Community, introducing in detail the planning and design methods of residential area, which respect nature and utilize the ecological potential of the site (Simonds, 1997).

In China, there have been philosophical thoughts of symbiosis with the natural ecology for a long time, such as "nature-human integration." These eco-philosophical ideas have a certain influence on China's development, including China's urban construction. In instance, classical garden design and urban "Feng Shui" theory are inextricably linked with nature. In modern times, Academician Wu put forward the theoretical framework of "Introduction to Sciences of Human Settlements" (Wu, 2001) based on the concept of generalized architecture, which is a trinity of architecture, landscape, and urban planning, laying a theoretical foundation for urban ecological design.

Later, from the perspective of urban planning, some scholars put forward urban ecological planning methods, using spatial ecologicalization as the core, and systematically explored ecological space planning and design methods applicable to different scales and levels (Xing, 2007; Yang, 2005). At the same time, some scholars started with the small-scale environment and proposed the design method of "basic ecological unit of human settlement" (He, 2005). At present, China has achieved certain achievements in the construction of ecological cities, but its focus is limited to ecological design from scratch, and the scale is either at the city level or at the unit level. It lacks the architectural environment design theory to solve the existing urban ecological problems.

In summary, the theory and practice of ecological restoration of built environment have achieved certain results. Multidisciplinary research, on the basis of built environmental science and traditional restoration ecology, has promoted the development of ecological restoration theory and has also contributed to the relevant practice.

2.3   How to Achieve Ecological Restoration in Architectural Design

2.3.1   Ecological Characteristics of High-Density Urban Environment

In high-density urban areas, the external environment of the building and the internal environment of the building together constitute the high-density urban environment. The external environment of high-density buildings, as the connecting part of the artificial building environment and the natural environment, is affected by both the built environment and the natural environment and has the ecological characteristics of complexity, openness, and instability.

2.3.1.1   Complexity

Human living behavior will inevitably involve the transformation of nature. The initial living style such as burrowing and nesting reflected the minor transformation of the natural environment by humans. With the improvement of science technology and the progress of engineering technology, humans can transform nature on a large scale in various ways to adapt to their own lives, including reorganizing natural elements to form a high-density urban environment. Humans control the circulation of material and energy in the high-density urban environmental ecosystem through technological means, thereby transforming and forming an environment consistent with aesthetic concepts and lifestyles. Therefore, high-density urban environmental systems often have the dual characteristics of natural ecology and humanistic spirit.

From the perspective of the constituent elements, in addition to natural elements such as sound, light, heat, plants, etc., the high-density urban environment also includes artificial elements such as road traffic facilities, environmental landscape facilities, public venues, etc., so as to meet the citizens' physical, psychological, and cultural demand. The diversity of constituent elements and the particularity of satisfying functions determine the complexity of the high-density urban environment.

2.3.1.2 Openness

The high-density urban environment is not a completely independent environmental system. It exchanges material, energy, and information with the natural environment, leading it to be an open system. Although some urban areas are relatively independent and are a comprehensive "society-economy-natural" complex ecosystem, it still requires extensive exchange with the natural environment to maintain the system's material and energy constant. High-density urban areas are the conversion nodes for natural resource input and waste output. The subject of energy conversion is humans. Humans convert natural resources into sewage, garbage, and waste gas through their daily behaviors, meeting their own development needs through this conversion, which includes life conditions, spiritual and cultural requirements. The process of energy circulation is ensured by this manual maintenance as well. In addition, the high-density urban environment obtains temperature and humidity that meet human needs through energy exchange, thereby ensuring its sustainable development. The openness of the high-density urban environment makes the development of nature and urban synchronize and influence each other.

2.3.1.3 Instability

It is undeniable that the improper transformation of high-density urban environment by humans will break the original ecosystem, destroy the balance of the biological chain, and reduce the biodiversity. The organisms in natural ecosystems tend to evolve toward the diversification of natural evolution and can remain relatively stable under external influence; however, for an environment dominated by artificial ecosystems such as high-density urban areas, if human influence and intervention exceed the regulation capacity of the ecosystem, it is extremely easy to destroy the ecological balance of the external environment of the settlement. In ecology, the capacity of the system is generally measured by "environmental capacity" (Wei, 2011). Once the load exceeds its environmental capacity, it will cause continuous damage to the ecosystem, and it will be difficult to recover through self-repair capabilities.

2.3.2 Filter and Determine Ecological Index Based on Local Environmental Characteristics

Determining ecological index is the basis for establishing a target system for ecological restoration of high-density urban environment. At present, many countries have established building evaluation index systems that include some ecological environment elements, such as the United Kingdom, the United States, Japan, Singapore, and China. The LEED-CC certification system in the United States, the

Table 2.1 Ecological restoration index system for high-density urban areas in Shanghai

Main criteria index		Sub-criteria index	
A	Greening environment	A1	Greenbelt system connectivity
		A2	Green coverage
		A3	Plant excellence rate
		A4	Proportion of multi-layer planting communities
B	Water environment	B1	Proportion of permeable pavement
		B2	Rainwater collection utilization rate
		B3	Water-saving rate for irrigation
		B4	Self-purification capacity of water features
C	Physic environment	C1	Noise insulation and reduction
		C2	Outdoor noise level
		C3	Road lighting effect
		C4	Daylighting effect
		C5	Shading rate
		C6	Material reflectivity
		C7	Winter wind speed
D	Road and traffic	D1	Accessibility
		D2	Traffic flow
		D3	Parking space scale
E	Spatial environment	E1	Fitness area
		E2	Animal habitat
		E3	Public space accessibility
		E4	Corner space utilization
F	Waste discharge	F1	Odor emission
		F2	Pollutant diffusion
		F3	Garbage collection
		F4	Organic waste biochemical treatment

BREEAM certification system in the United Kingdom, and the "Green Building Evaluation Standards" in China are widely used. It should be noted that ecological index has strong locality. In practical applications, appropriate ecological index should be formulated with reference to existing evaluation systems as well as local conditions. Here we take Shanghai, China, as an example, to provide a sample of the ecological restoration index system for high-density urban environment (Table 2.1).

2.3.3   Analyze the Cause of Damage and Measure the Degree of Damage

The causes of ecological degradation of built environment are complex, diverse, and dynamic. Natural conditions and human activities are the main causes of degradation. The foundation of the research on ecological restoration of built environment is to grasp the reasons that drive the degradation of the ecosystem and sort out the pressure on the ecosystem caused by different land use forms.

For the ecological restoration of the high-density urban environment, all the driving factors must first be properly analyzed; then the main influencing factors that cause the ecological damage or degradation of the high-density urban environment should be identified, so as to provide certain guidance for its ecological restoration practice.

In the process of ecological restoration of natural ecosystems, the "benchmark inventory" is usually used to describe the current status of the ecosystem (Clewell & Aronson, 2014). Through the investigation of the benchmark inventory, we can determine the cause of damage to the ecosystem, the extent of damage, and the existing stress factors and lay the foundation for the next step of ecological restoration.

The benchmark inventory survey is to describe the degree of damage to the ecosystem, not to describe the site comprehensively. There is no need to record content irrelevant to the purpose of the survey, while those key factors affecting the ecosystem worth more attention, on account that any slight change of them may causes damage to the ecosystem. Therefore, being aware of these key factors is crucial to formulating an ecological restoration strategy. The following table (Table 2.2) depicts the content of the natural ecosystem benchmark inventory report.

Similar to the method of using benchmark inventory surveys to measure the degree of ecological damage to natural ecosystems, the development of environmental and ecological restoration work in high-density urban areas requires an in-depth investigation of the environmental and ecological status of high-density urban areas. Firstly, a typical area should be selected, and then the ecological damage of the typical area's environment should be described in detail through on-site observation, spot testing, questionnaire surveys, interview records, etc., to determine its damage elements, damage degree, and cause of damage; finally experts can develop targeted ecological restoration strategies.

2.3.4 Formulate Ecological Restoration Strategies

2.3.4.1 Improve the Green Environment

Greening environment is an important component of the environmental ecosystem. Besides the generally recognized functions like beautifying and relaxing, the greening system plays a very important role in improving the ecological environment, maintaining the health of the ecosystem, and adjusting the microclimate. The ecological efficiency of greening is closely related to the design of vegetation communities. Generally speaking, the more reasonable the plant community structure and the larger the green area, the stronger its ecological efficiency is. To improve the urban greening environment and optimize the ecological efficiency of greening, it is

Table 2.2 The content of the natural ecosystem benchmark inventory report

Category	Contents	Methods
1 Species composition and community structure at the project site	Animals and plants observed on the project site, and judge whether these animals and plants contribute to the restoration of the ecosystem	Photography CAM GIS
	Species that need to be eliminated or controlled	
	Species that need to be introduced into the ecosystem	
2 Environmental status such as soil and hydrological conditions	Soil status	
	Hydrological characteristics	
3 The factors that cause damage to the ecosystem and the extent of its impact on the ecosystem	Factors affecting material exchange and biological migration between the project site and its surrounding environment	
	External factors affecting ecosystem restoration	
4 The impact of surrounding landscape characteristics on the ecological function of the project site	Local environmental changes	
	Changes in regional use of land	
5 Comprehensive assessment of damage to ecosystem functions		

necessary to increase the green space rationally, coordinate planting scientifically, enrich green space functions, and connect them to the green space system.

2.3.4.2 Reshape the Water Environment

Water is the source of life, directly affecting the growth and development of organisms and the structure of biological communities. Its natural circulation process plays an important role in the regulation of climate and surface climate. High-density urban environment is facing problems such as a small percentage of permeable paving, low rainwater utilization, water waste caused by irrigation, and fewer water features. Designers can use methods such as constructing permeable surface, rainwater collection and utilization, plant water-saving optimization, and adding aquascape to reshape their water environment (Fig. 2.2).

2.3.4.3 Improve the Physical Environment

Sound, light, heat, and wind are ecological elements that belong to the physical environment of high-density urban areas, and these elements all affect the comfort of citizens' lives. The comfort of the physical environment is also an important factor for citizens to participate in outdoor activities. Especially in the current society where the proportion of elderly people is gradually increasing, the comfort of the physical environment has more important health significance. Aimed at the problems of high-density urban physical environment such as large noise interference, obstructed vision during the day, uneven lighting radiation at night, high reflection rate, low shading rate, and uncomfortable wind speed, four measures are proposed

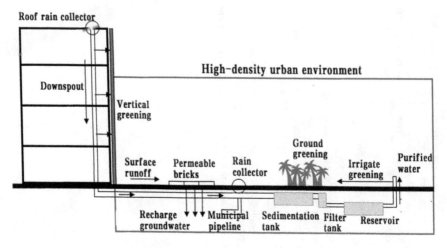

Fig. 2.2 Schematic diagram of rainwater recycling mode

to improve the physical environment of high-density urban areas, including isolating and reducing noise, constructing hard pavement shading, rearranging wind environment, and improving light environment.

2.3.4.4 Optimize Road Traffic

With the popularization of private cars, there is a serious shortage of parking spaces in urban areas, which is manifested by private cars occupying roads and destroying green spaces. The problem of parking difficulties has become increasingly prominent. In addition, high-density urban areas also have problems such as complex and chaotic traffic flow, crowded parking on the road, and lack of barrier-free systems. Road traffic in the urban area can be optimized by adjusting the traffic structure, establishing a flexible parking system, and road renovation. Improve the public transportation system within walking distance of citizens, raise parking fees, and other methods could guide citizens to adjust the transportation structure. The green space and parking spaces can be combined for restoration and design to form a shaded parking lot. The enclosed space formed by plants could be used to set up parking spaces, which can be used as a space for civil activities when there is no car parking, so as to realize the intensification of low green spaces in high-density environments use.

2.3.4.5 Reconstruct the Spatial Environment

The restoration of the external environment of high-density urban areas plays a key role in improving urban environmental vitality and promoting citizen interaction. For high-density urban areas with defects such as a single form of public space, insufficient fitness venues, lack of animal habitats, low accessibility of public spaces, insufficient vitality, and low utilization of corner spaces, we can make use of abundant venue space, symbiotic space creation, space vitality inspiration, and corner space utilization to reconstruct urban environment.

2.3.4.6 Regulate Waste Discharge

The collection, transportation, and treatment of garbage is an important link to ensure fresh air and clean environment in high-density urban environment. In view of the serious odor emissions in high-density urban areas, the low rate of classified collection and biochemical treatment of community garbage, it is high time that the garbage collection and processing chain should be established. Comprehensive restoration should be carried out from each link: single-household collection, garbage collection station, centralized transportation or treatment, using methods such as household garbage classified collection, and organic garbage reduction treatment (Fig. 2.3).

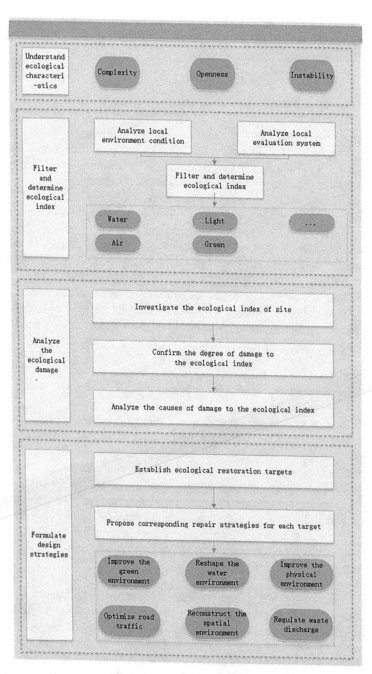

Fig. 2.3 How to achieve ecological restoration in architectural design

2.4   Case Study

2.4.1   Shanghai Anshan Eighth Village

Anshan Eight Village is one of the eight residential communities of Anshan New Village. The construction is completed in 1984, which means this living area is quite aged, leading to a poor environment design. According to the target system of ecological restoration of the high-density urban environment constructed above, the evaluation of the external environment of the target community can be divided into two aspects: quantitative and qualitative. The evaluation of quantitative indicators is mainly based on on-site investigations, the collection and sorting of external environment-related data, and the simulation by Honeybee, which involves plot ratio, parking space, greening rate, and other indicator data. The qualitative indicators are evaluated by questionnaire surveys of residents to obtain their evaluation opinions. Finally, the evaluation results are obtained through the fuzzy comprehensive evaluation method and the Likert evaluation quantitative standard, and the evaluation results are analyzed to summarize the ecological problems of the Shanghai high-density urban housing environment and then determine its ecological restoration strategy.

2.4.1.1   Greening Environment Evaluation

The greening environment mainly includes four aspects: the connectivity of the green space system, the coverage of greening, the rate of good plants, and the proportion of multi-layer planting communities. Through satellite maps, questionnaire surveys, on-site surveys, and other methods, it is found that the greening environment of Anshan Eighth Village is poorly connected, with a moderate green coverage rate (40%), a good rate of plant excellence, and a normal multi-layer planting community (Fig. 2.4).

Fig. 2.4 The greening environment of Anshan Eighth Village

2.4.1.2 Water Environment Evaluation

The water environment mainly includes the proportion of permeable pavement, rainwater collection utilization rate, irrigation water saving rate, and water feature self-purification ability. Through questionnaire surveys and on-site investigations, it is found that the current water environment in Eight Villages in Anshan is low percentage of permeable pavement, low rainwater collection utilization, low irrigation water-saving rate, and no water feature (Fig. 2.5).

2.4.1.3 Physical Environment Evaluation

The physical environment mainly includes noise reduction and isolation, road lighting effect, daylighting effect, shading rate, reflectivity, and winter wind speed. Through the use of TES-1341 Hot-Wire Anemometer and TES-1357 for on-site measurements, Honeybee for conducting simulation analysis of light environment and wind environment, and on-site investigations, it is found that the current physical environment of Anshan Eighth Village has poor noise isolation and reduction, relatively high outdoor noise levels (52.6 dB during the day, 44 dB at night), poor road lighting, good sunshine, average shading rate, poor reflectivity, and high wind speed in winter (0.2–1.8 m/s) (Figs. 2.6 and 2.7).

Fig. 2.5 The poor water environment of Anshan Eighth Village

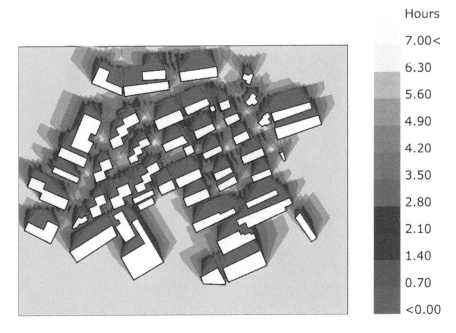

SunlightHours Analysis

Fig. 2.6 Sunlight analysis

2.4.1.4 Road Traffic Evaluation

Road traffic evaluation mainly includes barrier-free, traffic flow, and parking space scale. Through questionnaire surveys and on-site investigations, it was found that Anshan Eighth Village has defects of lacking barrier-free facilities, complicated traffic flow, and small parking spaces.

2.4.1.5 Space Environment Evaluation

Spatial environmental evaluation mainly includes the area of fitness venues, the existence of animal habitats, public space accessibility, and corner space utilization. Through questionnaire surveys and on-site investigations, it was found that Anshan Eighth Village has a small fitness venue (100 m²), fewer animal habitats, poor accessibility to public spaces, and low utilization of corner space.

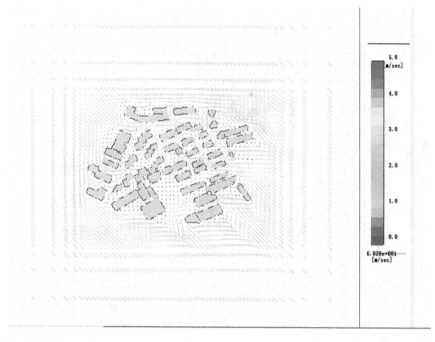

Fig. 2.7 Winter wind speed simulation

2.4.1.6 Waste Discharge Evaluation

Waste discharge evaluation mainly includes four aspects: odor discharge, pollutant diffusion degree, waste classification collection rate, and organic waste biochemical treatment degree. Through on-site investigations, it was found that the eight villages in Anshan have serious odor emission problems, average pollutant diffusion, good garbage collection rates, and no organic garbage treatment equipment

2.4.1.7 Ecological Restoration Strategies

Based on the above evaluation, the following Table 2.3 summarizes the ecological restoration strategies applicable to the residential area in the high-density urban environment of Shanghai:

Table 2.3 the applicable ecological restoration strategies

Category	Strategy	Highly recommended	Recommended	Not recommended
Optimize the green environment	Reasonably increase greening	●		
	Scientifically arrange planting		●	
	Rich green space function		●	
	Connect to the green space system		●	
Reshape the water environment	Construct the penetration surface	●		
	Collect and utilize rain	●		
	Optimize plant water-saving		●	
	Add water features			●
Improve the physical environment	Isolate or reduce noise		●	
	Add hard floor shading		●	
	Create the wind environment	●		
	Improve the light environment	●		
Optimize road and traffic	Adjust travel traffic structure		●	
	Build a flexible parking system		●	
	Renovate road facilities	●		
Reconstruct the spatial environment	Rich venue space	●		
	Create symbiosis space	●		
	Cultivate spatial vitality		●	
	Utilize corner space			●
Regulate waste discharge	Separately collect household waste			●
	Simplify the disposal of organic waste		●	

2.4.2 Meridian Water Regeneration

Meridian Water was originally an industrial and retail site in the northern part of the United Kingdom. It was home to many large buildings such as IKEA, TESCO, and Coca Cola factories. Since 2013, led by Enfield Council, designed by LDA, the regeneration program of Meridian Water area has been initiated. This program will bring 10,000 homes, a brand-new railway station, and thousands of jobs to Enfield. There are three pillars in this program: (1) Mixing uses and animating streets, (2) Park life at your doorstep, (3) Your place to make and create.

2.4.2.1 Enabling Movement and Improving Transport Connection

In the new reconstruction plan, the dominance of private cars in this community has been lowered and it turned to support sustainable modes of travel, such as circle parking, electric cars, and car clubs. At the same time, in order to create a community with safe traffic, effective connections, and high accessibility, based on the existing public transportation and infrastructure, LDA integrates public transportation around the upgraded railway station in the master plan and combines it with bicycles and sidewalks, to form a new sustainable street transportation network.

2.4.2.2 Maximizing the Potential of the Waterways

There are several waterways and culverts flowing through Meridian Water, including River Lee, Salmons Brook, and Pymmes Brook. In collaboration with Lee Valley Regional Park, those water shape the natural feature of Meridian Water.

The master plan of Meridian Water tries to dig out the potential of the water. To maximize the earnings from Lee Valley's view, waterside platforms are created. Meanwhile, the reservoir and other waterways are designed to be accessible, which may promote leisure and recreation opportunities, offering residents, visitors, and workers a better ecological environment to live in.

2.4.2.3 Improving Access to Healthy Living Corridors

As mentioned, the nearby Lee Valley Regional Park provides significant ecological benefits to Meridian Water. In the regeneration plan, the importance of utilizing this park's ecology and biodiversity value has been highly emphasized. Therefore, the connections of east-west and north-south have been improved to be a network of open spaces, leading to a healthier living in this area and helping to reconnect surrounding communities.

2.4.2.4 Delivering High Levels of Sustainability

Meridian Water is targeted to be net zero carbon by 2030 and to meet other standards of sustainability such as The Code for Sustainable Homes standards (CfSH) for residential buildings and BREEAM excellence standards for others. New technologies are applied to reduce CO2 emissions, turn waste into a local resource, reduce water demand, manage flood risk, etc. Meridian Water attempts to promote the waterways and raise the biodiversity in this area, creating attractive and sustainable neighborhoods.

2.5 Conclusion

A good ecological environment is the material basis for human sustainable development. It is not only the material carrier for human life, but also the material carrier for the survival of all other organisms, and the survival of these other organisms will in turn have an impact on human life. In this case, restoring and maintaining the existing ecological environment is an inevitable requirement for humanity as a whole. It has been nearly 50 years since the concept of "ecological restoration" was put forward, and its introduction into architecture also has a history of more than 10 years. During this period of time, various disciplines are trying to carry out ecological restoration in the human settlement environment, especially for the high-density environment, which is more artificially disturbed. High density is the (Aflaki et al., 2017) unavoidable development intention of modern cities, but the destruction of the ecological environment by this highly artificial development intention can be avoided, and even the restoration of the ecological environment can be achieved. This article, which is supported by National Natural Science Foundation of China (Grant No. 51778424), summarizes the ecological restoration strategy system that architects can adopt in the face of complex high-density urban environments, from screening indicators, evaluating the environment to selecting strategies.

The strategy of ecological restoration is not a supplementary measure after the completion of the building, but should be the concept of ecological restoration throughout the design process. In 2020, human beings encountered a major infectious disease that spread rapidly. Obviously, our current environment cannot alleviate this spread. Facing the question that how to ensure the ecological environment can protect human health and sustainable development, what kind of answer should the architect submit?

References

Aflaki, A., Mirnezhad, M., Ghaffarianhoseini, A., Ghaffarianhoseini, A., Omrany, H., Wang, Z., & Akbari, H. (2017). Urban heat island mitigation strategies: A state-of-the-art review on Kuala Lumpur, Singapore and Hong Kong. *Cities, 62*, 131–145.

Bradshaw, A., & Wong, M. (2003). *The restoration and management of derelict land: Modern approaches*. World Scientific.

Clewell, A., & Aronson, J. (2014). *Ecological restoration*. Island Press.

Fahy, F., & Cinnéide, Ó. M. (2008). Developing and testing an operational framework for assessing quality of life. *Environmental Impact Assessment Review, 28*(6), 366–379.

Forman, R. T., & Godron, M. (1986). *Landscape ecology*. Wiley.

He, Y. (2005). *Research on suitability of human settlement environment—Concept and method of "basic human settlement ecological unit"* (PhD thesis). Zhejiang University.

International Union of Architects. (1981). *The Warsaw declaration of architects*. International Union of Architects.

Jenks, M., & Jones, C. (Eds.). (2009). *Dimensions of the sustainable city*. Springer Science & Business Media.

Radhi, H., Fikry, F., & Sharples, S. (2013). Impacts of urbanisation on the thermal behaviour of new built up environments: A scoping study of the urban heat island in Bahrain. *Landscape and Urban Planning, 113*, 47–61.

Simonds, J. O. (1997). *Landscape architecture*. McGraw-Hill.

Society for Ecological Restoration International Science & Policy Working Group. (2004). *The SER International primer on ecological restoration*. Society for Ecological Restoration International. www.ser.org

Song, S. Q., & Zhou, Y. Z. (2001). *Mining wasteland and its ecological restoration and reconstruction*. Conservation & Utilization of Mineral Resources.

Wei, X. (2011). Discussion on environmental ecological planning of urban residential district. *China Science & Technology Panorama Magazine, 2011*(13), 259.

Wu, L. (2001). *Introduction to sciences of human settlements* (1st ed.). China Architecture & Building Press.

Xing, Z. (2007). *Marginal zones and marginal effects: A broad urban and rural ecological planning perspective* (1st ed.). Science Press.

Xue, M., & Luo, Y. (2015). Dynamic variations in ecosystem service value and sustainability of urban system: A case study for Tianjin city, China. *Cities, 46*, 85–93.

Yang, P. (2005). *Research on the theory and method of urban and rural spatial ecological planning* (1st ed.). Science Press.

Yang, Y. (2008). A tale of two cities: Physical form and neighborhood satisfaction in metropolitan Portland and Charlotte. *Journal of the American Planning Association, 74*(3), 307–323.

Yanitsky, O. N. (1987). Social problems of man's environment: Where we live and work. *Cities and Human Ecology, 1*, 174.

Zhao, J., Dai, D., Lin, T., & Tang, L. (2010). Rapid urbanisation, ecological effects and sustainable city construction in Xiamen. *International Journal of Sustainable Development & World Ecology, 17*(4), 271–272.

Chapter 3
Urban Public Landscape Design Under the Guidance of Sustainable Stormwater Management

Chunxiao Wang

3.1 Introduction

Luna B. Leopold, an American geomorphologist and hydrologist, once said, "Of all land-use changes affecting the hydrology of an area, urbanization is by far the most forceful." Throughout human history, water has remained one of the most important elements in urban development. However, the natural hydrological environment has been drastically altered in highly urbanized areas, and the properties of underlying surfaces have changed. Large areas of impervious buildings and surfaces have replaced the natural hydrological features. Consequently, the capacity of water conservation in cities has significantly reduced, and cities are vulnerable to the threat of floods caused by rainstorms. Floods have caused more damage and loss of life than any other natural disaster in the Asia-Pacific region. In Asia, for instance, there are nearly 2000 floods between 1995 and 2015, killing 332,000 people and affecting another 3.7 billion. However, stormwater management developments are mostly in developed countries, in most developed countries it's always a problem which is "nor that urgent," which makes the issue more challenging in both national polices and techniques.

Presently, there are two main reasons for the occurrence of flood disaster in cities: one is the changes in the properties of urban underlying surfaces, and the other is the loss of natural watercourses. In a natural rainwater circulation system, nearly half of the rainwater is absorbed by the ground surface, 40% evaporates via transpiration, and the remaining 10% is retained to generate surface runoff. In urban built-up areas—where more than 70% of the land surface is typically composed of impervious gray infrastructure—only 15% of rainwater can penetrate the surface,

C. Wang (✉)
School of Architecture and Urban Planning, Shenzhen University, Shenzhen, China
e-mail: cxw@szu.edu.cn

30% of rainwater evaporates via transpiration, and 55% becomes surface runoff (FISRWG, 1998). Furthermore, most early cities were built to have combined sewer systems (CSSs); that is, they feature sewers designed to discharge domestic sewage and stormwater in the same pipes. CSSs help in reducing engineering construction costs but can cause numerous environmental problems. When inlets are obstructed or pipe capacities are exceeded, stormwater will flood streets or building basements. In worst-case scenarios, the domestic sewage and stormwater mixture can overflow, causing water-pollution problems that cannot be ignored.

Therefore, sustainable management and utilization of rainwater is an inevitable trend in current urban constructions. At present, stormwater management in many countries has shifted from the traditional engineering drainage to ecologically sustainable stormwater management, which combines both infrastructure and ecological design. This approach makes full use of the natural ecological water circulation system to conserve water resources, protect water quality protection, mitigate pollution, and utilize water resources to realize sustainable urban development. Urban public landscape is not only an aesthetic and symbolic space, but also an ecological conduit and channel. Public landscape should help to enhance the resilience of cities in adapting to environmental changes and natural disasters. It should allow the rainwater to be absorbed, stored, infiltrated, and purified so that water can be released and utilized when needed, thereby enhancing the function of the urban ecosystem and reduce the occurrence of urban flood disaster and water pollution.

3.2 Overview of the Theory and Practice of Sustainable Stormwater Management

3.2.1 Theoretical Developments in Sustainable Stormwater Management

Stormwater management system is a set of related concepts, theories and technologies, laws and policies, and management mechanisms accumulated through long-term exploration, including both research and practice. Before the 1970s, stormwater was managed mainly by engineering approaches, including the construction of pipes and canals, storage ponds, and retention ponds. Since the 1970s, some countries have developed relatively perfect models and methods for sustainable stormwater management and introduced country-wide or city-wide ecologically sustainable stormwater infrastructure management systems. In 1972, the Federal Water Pollution Control Act Amendment proposed the concept of best management practices (BMPs) for the first time, focusing on the pollution control of non-point source water, which has been used widely in the United States and Canada. In 1990, the state of Maryland proposed the concept and technology system of low impact development (LID). The core concept advocates managing rainwater at the source location through a decentralized and small-scale approach (Sopes, 2010; US EPA,

2000). Water resources are permeated, filtered, evaporated, stored, and utilized through a series of engineering approaches (Sopes, 2010; US EPA, 2000; PGDER, 1999). LID is a microscale stormwater management technology developed through control measures. Compared with the traditional stormwater infrastructure, the LID concept emphasizes on construction on the basis of protecting and restoring the natural hydrological characteristics of the site with lower construction costs and more diversified ecological benefits.

The GI Working Group, organized by the American Conservation Foundation and the USDA Forest Service in 1999, developed the concept of Green Infrastructure, which is defined as "natural life support systems" comprising an interconnected network of waterways, greenways, wetlands, parks, forests, farms, and other protected areas that maintain the ecosystem and improve people's quality of life. Green infrastructure is a multifunctional open-space network at the urban and regional scales. It is defined as an approach to stormwater management that mimics natural hydrological processes at the local and site scales. In terms of the influencing scope, GI focuses on the integrity of the urban open space, whereas LID emphasizes microscale control methods for stormwater management. In terms of content, GI includes green space patches, connecting corridors, and natural or artificial urban green space networks, whereas LID focuses on more specific engineering measures, including bio-retention ponds, rainwater filtration systems, green roofs, rain gardens, infiltration trenches, and stormwater collection devices.

These theories were developed to address different urban water environment problems, and they have slightly different focuses. However, the common core aim is to resolve urban stormwater and water pollution problems through natural, sustainable means and to construct a benign urban hydrological cycle (Table 3.1).

3.2.2 History of Sustainable Stormwater Management in Practice

During the past two decades, many countries and cities have implemented several related practices based on the theoretical achievements represented by best management measures, LID, and green infrastructure. These efforts have increased in scale for the systematic sustainable stormwater management and construction in cities.

Table 3.1 Comparison of GI and LID

	Green infrastructure (GI)	Low impact development (LID)
Scale	Macro, meso, micro	Micro
Research focus	Integrity of urban open space	Microscale control methods for stormwater management
Research object	Green space patches, connecting corridors, and networks of natural or artificial urban green spaces	Bio-retention ponds, rainwater filtration systems, green roofs, rainwater gardens, infiltration trenches and rainwater collection devices, etc.

For example, the UK developed the Sustainable Urban Drainage System (SUDS), Australia proposed the Water Sensitive Urban Design (WSUD), New Zealand put forward the Low Impact Urban Design and Development (LIUDD), Seattle Public Utilities implemented the LID program, and Singapore developed Active, Beautiful, Clean Waters Design Guidelines (ABC Waters).

In recent years, China has developed several practices aimed at ecologically sustainable stormwater management, called "sponge city" construction. The term "sponge city" refers to the capability of a city to act like a sponge and be resilient to environmental changes and natural disasters caused by rainwater, which is also called water resilience city. In October 2014, the Ministry of Housing and Urban-Rural Development issued the "sponge city construction technical guidelines—low-impact development of stormwater system construction" and started the attempt to combine sustainable stormwater management and urban master planning. The "Green Building Assessing Standards" issued in 2014 in China also introduced the concept of LID, put forward overall planning requirements for stormwater control, and conducted quantitative evaluation of technical stormwater manage measures in combination with evaluation indicators. In a word, stormwater management in China has gradually shifted from engineering management to sustainable ecological management (Table 3.2).

In short, many countries and cities have adopted many sustainable stormwater management practices based on the theoretical achievements represented by BMPs and LID on a large scale for systematic sustainable stormwater management in cities. Now, the research direction in urban ecological rainwater management has gradually shifted from the hydrological control method with focus on runoff and water quality to a more comprehensive planning system combining urban landscape with infrastructure. Stormwater management now focuses more on the organic integration of engineering measures with the urban landscape while emphasizing the stormwater utilization and ecological benefits of GI.

3.3 Sustainable Stormwater Management-Oriented Urban Public Landscape Practices

The concept of sustainable stormwater management has been put forward during the past 20 years, whereas the concept of ecologically sustainable stormwater management has existed for a longer time. For example, Back Bay Fens in the Boston Park system planned by Olmsted includes a hydrological and rainstorm management system (Corner, 2006), representing an early practice of urban sustainable stormwater management. Since the 1990s, there has been a tendency to combine sustainable stormwater management with urban public landscape, and extensive practical studies have been performed in many countries. At present, depending on the scale and morphology, urban public landscape for sustainable stormwater management can be classified into the following categories: linear urban spaces, small

Table 3.2 Practice of sustainable stormwater management

Proposed time	Proposer	Management model/ methodology	Ideas
1970s	US Federal Water Pollution Control Act Amendments	Best Management Practices (BMPs)	Prevention or control of non-point source pollution and assurance of water quality through single or multiple BMPs
1990s	Maryland, USA	Low Impact Development (LID)	Emphasis on maintaining and protecting the natural hydrological functions of the site through small, decentralized control facilities at the source location
Late 1990s	Australia	Water Sensitive Urban Design (WSUD)	Integration of the hydrological cycle with urban planning, design, and construction development processes to reduce the need for structural measures through rational design
1996	Supplementary provisions of the German Federal Water Act	Mulden Rigolen System, MR	Emphasis on "zero drainage growth," sustainable rainwater collection and utilization
2004	UK National Working Group on Sustainable Drainage Systems	Sustainable Urban Drainage System (SUDS)	Management and preventive measures, source control, site control and area control
2006–2013	New Zealand	Low Impact Urban Design and Development (LIUDD)	Integration of multiple concepts to minimize the negative effects of urbanization by following the material cycles and energy flows of the natural ecosystem
2009–2018	Singapore	Active, Beautiful, Clean Waters (ABC Waters) Design Guidelines	To realize the full potential of this water infrastructure by integrating the drains, canals and reservoirs with the surrounding environment in a holistic way.
2012	China	Sponge City Construction	Cities can work like a sponge and have "resilience" in adapting to environmental changes and responding to natural disasters. When it rains, water can be absorbed, stored, infiltrated, and purified. When needed, water can be released and utilized
2014	China	Green Building Assessing Standard	Manage the site stormwater runoff, and control the total amount of stormwater discharged from the site. Recycle stormwater for irrigation or toilet flushing.

Table 3.3 Public landscape design strategy guided by sustainable stormwater management

Type of space	Type of public landscape	Design strategy and approach
Urban linear space	City streets, city waterways, walkways, alleys	Constructing ecologically sustainable rainwater collection ponds and converting existing road curbs into an open pattern that allows stormwater to enter the road green space. Completing the collection, mitigation, purification, and infiltration of stormwater from the street while retaining its original functions
Small, inward spaces	Parking lots	Striped planting areas, rainwater infiltration strips (to temporarily store and purify rainwater), tree plantings, permeable pavement, sand filter layer, and green roofs on garages
	Public buildings and ancillary grounds	The government and relevant authorities should provide policy support and guidance for self-construction; measures include green roofs, permeable pavement, and fee-for-service strategies for stormwater management
	Residential land	Rain gardens, green roofs, rainwater collection barrels, permeable pavement, etc. that integrate with the terrain. Collect rainwater and serve an educational purpose for the local community
Large outward spaces	Industrial land, public service land	A nonprofit organization that assesses the green potential of public facilities and encourages public oversight and integration with green streets
	Green spaces and plaza sites	Permeable pavement and green space. City parks and green spaces have relatively less impervious areas; however, they have significant potential to manage stormwater from surrounding streets

inward spaces, and large outward spaces. Their corresponding planning and design strategies differ (Table 3.3).

3.3.1 Urban Streetscape

Urban roads make up most of the linear space of modern cities. Roads have poor permeability and are often lower than other interfaces in the city. Therefore, rainwater accumulates on roads when the rainfall is significant. In addition, owing to the vehicle emissions and fuel leaks, roads are also a major source of runoff pollution on rainy days. Currently, urban road rainwater drainage depends mostly on the traditional drainage pipes. When the rainfall is heavy, water inlets can be easily blocked and/or the water quantity exceeds the carrying capacity of the drainage pipes. Therefore, rainwater will accumulate on the road surface, causing serious water pollution and flooding problems.

Common solutions to stormwater problems in urban linear spaces are retrofitting of the existing linear plantings with ribbon, decentralized, rainwater-absorbing bioretention basins, stormwater planting ponds, vegetated shallow trenches, and permeable pavements. Stormwater ponds are often composed of natural landscape elements such as plants, soil, sand, and gravel and replace the conventional above-grade ponds with planted areas with no tracts or openings to form ribbons of natural road stormwater management systems that retain stormwater, reduce runoff, promote infiltration of stormwater, and purify polluted water.

Portland is the second largest city on the West Coast of the United States after Seattle. As one of the pioneer cities to build ecologically sustainable stormwater streets, Portland has always been at the forefront of green street construction and has produced several excellent design examples. The SW Montgomery Green Street project is an example. The project covered several areas in and around the city with a "curbless" streetscape. The "curbless street" concept presented an approach to stormwater utilization in a highly urbanized city center. It not only included sustainable stormwater management but also maximized the protection, creation, and integration of the urban pedestrian space. The project won the American Society of Landscape Architects (ASLA) Award for Analysis and Planning in 2012 (ASLA, 2012).

In the SW 12th Avenue Green Street Project in the city of Portland, four "rainwater planters" were redesigned to follow the slope of the roadway using the green belt between the pedestrian and vehicular routes. Each rainwater planter is bounded by a precast concrete slab and planted with moisture and drought tolerant plants such as Juncus patens and Nyssa sylvatica (ASLA, 2006). Plants in the rainwater planters can slow down runoff, effectively purify stormwater containing impurities and sediments, and promote natural infiltration of rainwater, thus forming a natural management system for road stormwater. When the rainfall is heavy, water in the first rainwater planter overflows into the remaining planters. When the capacity of all planters reaches saturation, the overflow rainwater is discharged into the municipal drainage system. The design significantly reduces the net flow of stormwater from city streets, relieving pressure on the city's drainage infrastructure and reducing the potential for water pollution. Experimental data showed that the ecologically sustainable stormwater infrastructure project on SW 12th Street was able to reduce the runoff intensity of a 25-year rainstorm by at least 70%.

Owing to the limitations of the land use in city centers, planners began to pay attention to the integration of the various functions of streets for stormwater management. For example, Olin's plan of "Meeting Green" for Philadelphia proposed several innovative ecological planting pools combined with urban public facilities such as a theater, seats, and parking. The integration not only reduced the street space occupied by rainwater planters but also turned the space into a variety of infrastructure elements to serve pedestrians, thereby maximizing the use of space while creating a beautiful and practical urban landscape. It became part of the city's public space and gave the direction for the future sustainable development of urban communities.

3.3.2 Green Buildings

Structures occupy most of the surface area in highly urbanized areas. Urban storm-water runoff can be reduced by more than 60% if rainwater can be effectively managed on the roofs of structures and buildings. In countries with water shortages, stormwater is more economic friendly and suitable for seasonal use, such as for irrigation, landscape water, cooling, and other seasonal purposes. Meanwhile, stormwater storage pools can also be used as emergency water sources. In regions precipitation are balanced throughout the year, stormwater is more suitable for non-seasonal use like toilet flushing. The vertical design of the site should be conducive to the collection and discharge of stormwater, to organize the infiltration, retention or reuse of stormwater more effectively.

In terms of index control, for example, the "Green Building Assessing Standard" in China brought in the implementation of the outward discharge stormwater amount control. For sites larger than $10hm^2$, special design of stormwater control and utilization are required; for sites smaller than $10hm^2$, detailed designs are not essential, but stormwater utilization measures are encouraged. Specific evaluation indexes include number of bioretention, the proportion rain gardens in total area, the proportion of permeable pavement area in hard paved ground, etc. At present, the roof garden is a relatively successful method to combine the ecological rainwater infrastructure with the structure. By arranging plants on the roof to create land-scape, the falling rainwater is managed at the source location, which can effectively utilize rainwater resources, increase the ecological benefits of structures, reduce the surface temperature of buildings, and improve the overall ecological environment of the city. The rainwater collection cycle includes interception, diversion, and termi-nal storage. Through the preliminary interception by way of roof gardens and other measures, the remaining rainwater is dredged to the terminal storage space or into urban drainage pipes. As the most important part of the preliminary interception, the design of roof gardens has two key points: one is that the weight of the roof soil and plants should not exceed the load limit of the structure; the other is that engineering measures for waterproofing and seepage prevention must be handled carefully.

3.3.3 City Squares and Open Spaces

Open urban public space often includes a large area of hard pavements. When the rainstorm comes, it produces site ponding and a significant amount of runoff, which causes inconvenience to people and can disrupt their daily life. In public landscape design, stormwater management should be embedded in the design on the premise of meeting the traffic, use, and function demands. At present, there are three main methods for sustainable rainwater management in public spaces: setting up rain gardens based on site conditions, using permeable pavements, minimizing impervi-ous areas, and using rainwater to create multifunctional flexible landscape. In a site dominated by large areas of pavements, minimizing the area of impervious surfaces is the most direct way to control stormwater runoff. Solutions include increasing green areas and using permeable pavements. However, the reconstruction cost is high and the ecological benefit is limited. A rainwater garden can be set up based on the specific needs of the site, and the rainwater can be filtered and reused to maximize the ecological benefits.

3.3.4 Case Study

3.3.4.1 Bishan-Ang Mo Kio Park, Bishan, Singapore

Located in the tropics, the average annual rainfall in Singapore is about 2400 mm. However, due to its limited land and complex geological conditions, water resources are in short supply in Singapore to some extent. At the same time, urban water demand is expected to double over the next 50 years due to population growth and economic growth. As the longest river in Singapore, the 10-km-long Kallang River runs through Central Island and is an important part of the city's water supply system. During the 1960s and 1970s, Singapore built a system of concrete drains and canals to prevent widespread flooding, several key points of the Kallang River were constructed as concrete channels to discharge stormwater quickly during the rainy season.

Located in the heart of Singapore, Bishan Ang Mo Kio Park is one of the most popular parks. Built in 1988, the park was originally designed to create a green buffer zone and to provide leisure space for recreation. However, the drainage pipe became like a clear dividing line between the park and community as a straight fenced concrete canal.

To solve these problems, Bishan-Ang Mo Kio Park was re-designed and constructed. The redevelopment provides a new model for tropical urban hydrology through the instrument of landscape infrastructure, addressing Singapore's dual need for water supply independence and flash flood management while creating access to a thriving riverine ecology within the dense city. One of the brave moves was to break the concrete canal and restore the river exceeded the targeted carrying capacity while costing 15% less than the redesigned concrete canal. Simple, yet highly engineered, this blurred line between park and river has transformed the community's pragmatic perception of urban water systems to a relationship that is proud and close to nature.

Encouragingly, the water storage capacity of the transformed ecological river has been greatly increased by 40%, biodiversity has been increased by 30%, and recreational space has been increased by at least 12% (ASLA, 2016). Instead of the traditional approach which is to control stormwater inside concrete canals, the soft and beautiful new river landscape brings water closer to the people with more potential open space for recreation. The river is now much closer to its natural form, with different widths, creating valuable, natural, and diverse habitats for wildlife communities (Figs. 3.1, 3.2, and 3.3).

Fig. 3.1 The Kallang River during the dry season. During the rainy season, it could swell and retain stormwater then slowly discharge it downstream

3.3.4.2 Vanke Architecture Research Center (VARC), Guangdong, China

Vanke Architecture Research Center (VARC), designed by Chinese landscape design firm Z + T STUDIO, is an excellent example of sustainable stormwater management in an industrial park. It won the 2014 ASLA Universal Design Honor Award. VARC, located in Guangdong, China, focuses on studies related to the housing industry. The project used low maintenance materials and plants to achieve sustainable stormwater management of the site. Through the ingenious design strategy, rainwater collection and utilization are enhanced quantitatively and qualitatively. In "Ripple Gardens," the designer increased the time of rainwater landing by using arbors, and designed a wavy lawn combined with a slope design to increase the time for rainfall infiltration, thus controlling the rainwater flow. In "Windmill Garden", a 32-m tall windmill provides power for pumping the collected stormwater to the building roof for aeration treatment. Next, the stormwater passes through a series of plant purification ponds. The purified water is used for the site and playgrounds, meeting the multiple needs of ecology, education, and entertainment.

Fig. 3.2 Wetland plants plays an important role in water purification

3.3.4.3   Canal Park, Washington, D.C., USA

Canal Park in Washington, D.C., was transformed from a bus parking lot into a vibrant public gathering place. Rainwater is collected from the roofs of surrounding commercial buildings and surrounding roads during rains and piped into the park's rainwater planters, where it is filtered and purified by plants. Depending on the level of purification, treated rainwater can be used for park fountains, waterscape, and toilet flushing in surrounding buildings. In winter, the rainwater fountain is transformed into an ice-skating rink, forming a unique seasonal use function. Post-completion surveys have indicated that the site infrastructure including the rain garden and rainwater planters can capture 2.84 million gallons of rainwater runoff per year, of which 850,000 gallon can seep into the soil through permeable pavements and ecological rainwater planters.

66% of the water needs of the park can be met by the collected rainwater. It can be used for fountains in summer, park irrigation, public facility cleaning, skating rink construction in winter, and washrooms (Figs. 3.4 and 3.5).

Fig. 3.3 Stormwater treated and purified by the wetland is reused in water facilities of children's play fields

3.4 Versatility of Sustainable Stormwater Landscape

The use of stormwater as a resource is meant to move away from the conventional approach of relying on underground drainage systems to collect, manage, and reuse the stormwater at the source location (Richman, 1999). In dense and arid cities, urban water infrastructure is under pressure. Therefore, rainwater, as a natural resource, should be efficiently collected and reused.

Compared with conventional urban drainage systems, the use of stormwater resources places greater emphasis on the integration of stormwater collection with services and recreation. One of the criteria for sustainable urban drainage design evaluation is the combination of the quantity and quality of collected rainwater and recreation (CIRIA, 2001). Therefore, the approach to stormwater utilization is to design sites that provide multiple functions and values. This versatility has the following features:

1. Practicality. Pressure can be relieved from urban gray infrastructure, and diverse habitats can be provided for plants and animals.
2. Education. Many rainwater gardens are built in schools. Along with the meeting the purpose of rainwater collection, rainwater gardens can help educate the public.

Fig. 3.4 Rainwater collected for activity in Washington Canal Park

3. Safety. The risk of flooding in the city can be reduced.
4. Recreation. Diversified urban open spaces and communication places can be provided for the public to experience the vitality of the landscape created by managing rainwater.
5. Aesthetics. Combined with artistic design techniques, a stormwater management approach can create an aesthetic context.

3.5   Conclusions

Stormwater resource utilization has become an important element in sustainable urban development. The ongoing, high-intensity urbanization visible across the globe has intensified the conflict between urban development and environmental protection. This is aggravated by water shortages, pollution, and floods. In China and many other similar countries where the spatial distribution of water resources is not balanced, but in the meantime, flood disasters occur frequently, it is of great strategic significance to plan a forward-looking rainwater resource utilization system. In recent years, China has promoted the popularization of the concept of sponge city, successful stormwater management planning and design works have been established, but the overall the development is still in the exploratory stage. From the past experience we can get the following enlightenment:

Fig. 3.5 Bioretention in
Canal Park

1. The importance of urban stormwater utilization must be further prioritized at the strategic level.
2. The multifunctional nature of urban ecological stormwater infrastructures must be explored.
3. Multi-departmental and cross-professional cooperation should be strengthened, especially mass participation.
4. Relevant quantitative indexes of legislative and regulatory bodies should be improved according to the local climate conditions, to promote the implementation of stormwater resource utilization at the policy level.

References

ASLA. (2006). *SW 12th Avenue Green Street Project*, Portland, Oregon [EB/OL]. http://asla.org/awards/2006/06winners/341.html

ASLA. (2012). *SW Montgomery Green Street: Connecting the West Hills to the Willamette River* [EB/OL]. http://www.asla.org/2012awards/572.html

ASLA. (2016). *Bishan-Ang Mo Kio Park*. https://www.asla.org/2016awards/169669.html

CIRIA. (2001). *Sustainable urban drainage systems: Best practice manual. Report C523.* Construction Industry Research and Information Association.

Corner, J. (2006). Terra Fluxus. In *The landscape urbanism reader* (pp. 9–11). Princeton Architectural Press.

Federal Interagency Stream Restoration Working Group (FISRWG). (1998). *Stream corridor restoration: Principles, processes and practice.* USDA. [EB/OL] (15 September 2006). http://www.nrcs.usda.gov/technical/stream-restoration

Prince George's County, Maryland. Department of Environmental Resources (PGDER). (1999). *Low-impact development design strategies: An integrated design approach* [EB/OL]. Document No. EPA 841-B-00-003. U.S. Environmental Protection Agency. http://www.epa.gov/owow/nps/lidnatl.pdf

Richman, T. (1999). *Start at the source: Design guidance manual for stormwater quality protection.* Bay Area Stormwater Management Agencies Association.

Sopes, J. L. (2010). *Sustainable solutions for water resources: Policies, planning, design, and implementation* (pp. 235–240). Wiley.

US EPA. (2000). *Low impact development (LID): A literature review.* EPA-841-B-00-005. United States Environmental Protection Agency.

Chapter 4
The Exploration of the Spatial Transformation of the Old Cities Under the Concept of Urban Repair Design: Take the Tianqiao Cultural Heritage Center as an Example

Qingguo Wang, Junjie Li, and Wenyan Bian

4.1 Introduction

"City Betterment and Ecological Restoration" is a "prescription" written out by the Ministry of Housing and Urban-Rural Development to treat "urban diseases," improve the quality of living environment, and transform the urban development mode, namely, ecological restoration and urban repair. The Central City Work Conference proposed to strengthen the urban design, promote urban repair, strengthen the planning and control of the spatial three-dimensionality, plane coordination, style integrity, and cultural continuity of the city, and retain the "genes" of the city, such as unique regional environment, cultural characteristics, architectural style, etc. (Ministry of Housing and Urban-Rural Development of the People's Republic of China, 2017). Nowadays, "Urban Double Repairs" has been highly valued, which plays a significant role in promoting the construction of public service facilities or green space, providing social open space and leisure space for citizens and improving the comprehensive environmental quality of a city. "Ecological restoration" pays more attention to the combination with the natural environment, while "urban repair" focuses more on the optimization of the urban environment (Yucheng & Xiang, 2020).

Q. Wang
China Architecture Design & Research Institute, Co. Ltd., Beijing, China

J. Li (✉) · W. Bian
School of Architecture and Design, Beijing Jiaotong University, Beijing, China
e-mail: lijunjie@bjtu.edu.cn

© The Author(s), under exclusive license to Springer Nature Switzerland AG 2021
S. S. Y. Lau et al. (eds.), *Design and Technological Applications in Sustainable Architecture*, Strategies for Sustainability, https://doi.org/10.1007/978-3-030-80034-5_4

The design scale of City Betterment is often uncertain. On the one hand, it refers to the method that exploiting systematic planning to integrate groups of large-scale buildings with the surrounding environment, which can activate this region's development. On the other hand, in old cities, it refers more to the method that making a small-scale single building embedded into existing groups of urban constructions based on their histories, which also means stitching cities incomplete place, in order to introduce new energy. City Betterment design is usually based on urban life to create, which is a process of shaping a space with heritage, memory, and vitality.

4.2   Spatial Transformation Methods of Old Cities Under the Concept of City Betterment

Urban restoration in the old city area has gradually transitioned from large-scale "demolition" to "organic renewal." City betterment is a sustainable development strategy for organic renewal. It does not mean that all old cities need to be repaired drastically. When we use this strategy to directly repair and refurbish historical blocks, its standard needs to be rethought.

Western architect and critic La Molesz once proposed "Urban Acupuncture" trying to combine the original texture of a city with the overall development of the city. They are not contradictory: "Urban Acupuncture" advocates that carry out small-scale interventions to improve the overall function of the city after selecting key "acupoints" (Zhongping, 2015). "Urban Acupuncture" emphasizes the large-scale "external influence" brought by small-scale renewal, which is a kind of superposition of quantity and is a gradual renewal focused on details that accumulates over time. However, the transformation of old urban areas in China is often the overall transformation of districts. The boundaries are relatively clear, and the updated content is more purposeful and accurate. Traditional Chinese medicine-style "Urban Acupuncture" does not have an immediate effect. Therefore, it is necessary to pay attention to combining Chinese and Western medical methods in the process of renewal. While preserving valuable old buildings and restoring and tracing the history, we should also darn the space that can inspire the vitality of the city and attractive public spaces and integrate the old buildings with the urban landscape, making the old and the new coexist.

"Elasticity" restoration takes advantage of its characteristics between the original function and the new function. It can use dynamic adjustment methods such as transition, communication, contact, and change to dynamically adjust according to needs in the process of urban sustainable development so that each functions organic symbiosis between bodies.

4.3   One Practice of Spatial Transformation of Old Cities Under the Concept of City Betterment

4.3.1   Location Analysis

The location of this project is in an area which represents a Beijing cultural symbol Tianqiao. Speaking of Tianqiao always reminds people of the traditional folk art of ancient Beijing. The site is in a shape of "L," facing Tianqiao South Street in the east and Beiwei Road in the south. In recent years, with the overall transformation of the performance space in the Tianqiao area, a group of high-quality theaters have been formed around the base. The southwest corner of the base is the Tianqiao Theater, the south side is the Tianqiao Performing Art Center and the Tianqiao Plaza, and the north side is the Capital Cinema. Although there are the headquarters of Deyunshe that everyone loves to hear, the birthplace of China Life, the Wansheng Theater, etc., its infrastructure cannot provide good services to the citizens. There are two problems:

The single units are chaotic and Deyunshe and Wansheng Theater cannot meet the needs of users in terms of function, form, structure, and fire protection.

The newly built buildings in the site are scattered, and the problem of unauthorized construction is serious. They have never responded to history in terms of form, scale, etc., have never respected the history, and have never tried to fit in the urban environment.

4.3.2   Search for the History and Trace to the Source

The traditional central axis of Beijing is the north-south axis of The Old City of Beijing, starting from Yongdingmen in the south and reaching the Bell and Drum Tower in the north. It is about 7.8 km long and has a profound impact on the urban development of Beijing. The central axis is the axis of history and culture, the soul and backbone of The Old City of Beijing, since the Yuan Dynasty, it has been the most significant benchmark and direction for urban development. It is also the crystallization of the wisdom of the Chinese people and significant cultural heritage.

According to a book in Qing Dynasty named "Guangxu Shuntian Fuzhi," "Yongdingmen Street is connected with Zhengyangmen Street in the north. There is a bridge named 'Tian'." ("Tian" means sky in English.) This bridge was used when the emperor went to the Temple of Heaven and the Temple of Xiannong to worship, so it is called a Tianqiao. In Chinese culture, people believed that the power of emperor is authorized by the sky. Tianqiao, the Temple of Heaven, and Xiannong Temple are arranged symmetrically on both sides of the Tianqiao. And these three constructions together form a significant part of the royal culture. The architect Sicheng Liang once said: "When you move forward north from the main gate on the south end, you can see two roughly symmetrical buildings, the Temple of Heaven and the Temple of Xiannong, located on the both sides of the central axis. When you pass through a long avenue opposite to the city buildings and arrive at the crossing of Zhushikou, you can see the first key point of the inner city, the majestic Zhengyang Gate Tower. In front of the gate about 100 m, there are a large archway and a large stone bridge blocking the road, which are regarded as an avant-garde for the first key point." Liang Sicheng's description clearly shows the sequence of Beijing's central axis, and the Tianqiao is one of the significant points on this axis.

The Beijing City Master Plan (2016–2030) proposes that Beijing is strategically regarded as a political center, cultural center, international exchange center, and technological innovation center. It clarifies the urban spatial structure including one main development area, one subsidiary development area, two axes, and multiple points. It also strengthens the protection, inheritance, and rational use of historical and cultural cities. In short, the central axis that runs through the history of Beijing's development has a profound impact on Beijing's urban development in terms of macro and culture. Currently, Beijing Central Axis is applying for World Cultural Heritage. And Beijing Tianqiao Performing Arts District, as an important area of the Central Axis, reflects the co-prosperity and symbiosis of royal culture and civilian culture. It is the cradle of Beijing folk art and a representative of Chinese folk culture.

At present, the performance space in the Tianqiao area is becoming more and more perfect. The central axis Royal Road landscape is restored following the historical texture. The building scale and form are relatively unified with the central axis landscape, forming a coordinated counterpoint relationship.

The urban renewal design should not only subtly conform to the urban atmosphere created by the central axis (solemnity, ritual sense), but also need to reflect the unique temperament of the Tianqiao culture.

4.3.3 Inherit the Function and Continue the Memory

With the development of history, the old Tianqiao has undergone tremendous changes, and people's memories of the Tianqiao have become unclear. But the three buildings, the Zhonghua Cinema, the only Tianqiao Cinema left, Deyunshe, and Wansheng Theater, all imply brilliant folk culture that once happened here. The Zhonghua Cinema has been rebuilt to meet today's needs in terms of form and function, but Wansheng Theater and Deyunshe cannot meet not met the requirements in terms of function, form, structure, and fire protection due to their age. On the south side of the project, The Beijing Tianqiao Performing Arts Center has been completed. The original intention of this design research is to supplement the theater types of the Performing Arts Center, integrate the Deyun Society, and expand the scale of the Wansheng Theater, in order to reproduce the prosperity of the old Tianqiao cultural market.

In the early years of the Republic of China, the Tianqiao area really became a prosperous civilian market, which was regarded as a typical area of the old Beijing civilian society. Here you can buy daily necessities and see all kinds of folk art and the high-quality and inexpensive flavored foods. These fun activities occurred in the original hutongs' texture. To reproduce the original sense of place is to respect the land (Jie, 2018).

Besides, there are some existing architectural facades that contain certain era characteristics and historical significance. For example, the Archway facade of Wansheng Theater and China Life's round arch facade (Figs. 4.1 and 4.2) will be integrated into the new functional space. In the renovation design of the Zhonghua Cinema on the north side of the base, the designers pay attention to the coordination and unity of street materials and scales. The newly built Wansheng Theater will also continue the volume of the Zhonghua Cinema and stitch the missing street interface.

4.3.4 Stitch the Streets and Fit the Shape

The facades of the buildings on both sides of Tianqiao South Street are relatively complete, with a height of about 18 m and 5–6 floors. The base is located on the north side of the Tianqiao square. The current architectural style, architectural form, and space scale are extremely inconsistent with the Beijing Tianqiao Performing Arts Center on the south side of the square, showing the lack and imbalance of the urban space (Fig. 4.3). According to the control requirements on the volume and

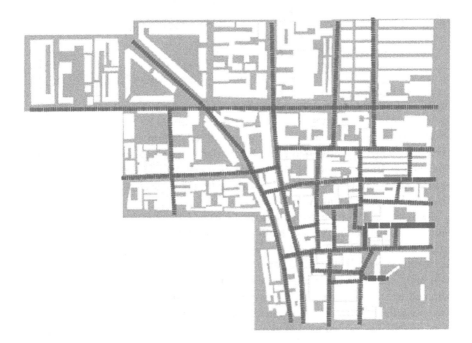

Fig. 4.1 The historical texture of the original streets and lanes in the base

Fig. 4.2 The facade of the historical building in the base

Fig. 4.3 The street facade of Tianqiao South Street

color of the buildings in the spatial sequence of the central axis, the new buildings of the base should strictly follow the requirements to supplement the street facade. And their proportions, scales and facade elements should echo the surrounding neighboring buildings uniformly handle the sidewalk scale along the street (Beijing Tianqiao Southern Acting Area, 2017).

The complexity of the design task is the excavation and sorting of history and culture, meanwhile the types of buildings surrounding the base greatly restrict the production of the plan. To meet the sunshine requirements of residential houses in the northwest corner of the base, strict sunshine calculations are required for newly built buildings. As a result, the shape of high in the south and low in the north is obtained as shown in the figure, sunshine simulation calculation through reverse reasoning, the building shape is calculated, and then fit the shape, and a complete slope interface is formed by (Fig. 4.4).

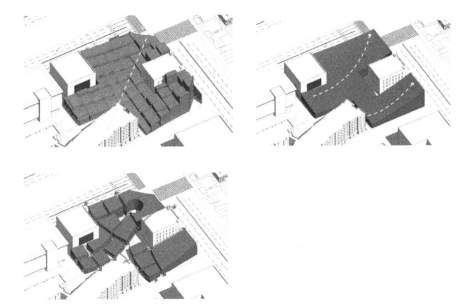

Fig. 4.4 The formation process of building shape

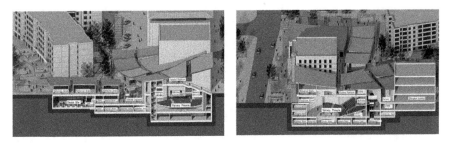

Fig. 4.5 Schematic diagram of building section

The restoration strategy attempts to find site clues from the original texture of the streets and hutongs in the old Tianqiao area. Then use these clues as design basis to re-divide the building units in the site and create small-scale architectural settlements on the ground floor that conform to the traditional space, rather than a collage of blocks with simple function. Commercial and exhibition spaces are placed on the ground floor, while large theater spaces are placed underground (Fig. 4.5). After the segmentation, the volume scale has become smaller, which provides more experience space for pedestrians and adding more exhibition interfaces.

4.3.5   Sharing of Neighborhoods and Community Integration

The theatrical building needs to be equipped with sufficient parking spaces. However, due to the unusual location, the space between the building and the urban roads is very narrow. And the surrounding roads do not have conditions to open a gate for the garage. Besides, the interior and underground spaces of the site cannot provide sufficient parking conditions. Therefore, an underground passage is set up to connect with the Beijing Tianqiao Performing Arts Center on the south side and the underground parking garage of the plaza. At the same time, the convenient underground and above-ground parking spaces around the project are reasonably used to meet the parking requirements. A small number of temporary parking spaces are set up along the North Latitude Road to facilitate the loading and unloading of props. The newly built buildings take not to aggravate surrounding traffic pressure as the basic principle. Through reasonable traffic and people flow organization, regional traffic gets a more positive impact.

The sharing of block resources solves the demand for parking resources. Also, the boundary effect produced by the different interfaces of the center can be fed back to the surrounding communities. For example, the low buildings on the north side will not affect the sunlight demand of the residences, and the sufficient retreat distance exposes the narrow and messy inner courtyard of the first floor of on the northern part of the base. By changing the surrounding walls of housing zone into fancy brick walls, which not only encloses the activity space inside the community but also forms effective isolation, it increases activity platforms, green spaces, and landscape sketches in the community to create a pleasant living environment. On the inner street side, some small-scale resting spaces are also set up along the wall. It can be seen that the newly built buildings no longer serve as a strong intervention but as a catalyst to soften boundaries and activate neighborhood relations (Jingqiu & Shifen, 2020).

To maintain the outward output of cultural performances, combine the diversity of venues, and serve as many surrounding residents as possible, the design uses the original Tianqiao archway on the west side as the main entrance of Deyunshe, and Deyunshe is placed underground, placing the community shared theater and a slow-moving system into the the large-scale Tianqiao Market Slanted street facing the west side of the base.

4.3.6   To Refine the Elements and Sculpt the Details

The Zhonghua Cinema and Tianqiao Art Building on the north side are three-section architecture, with roof, main body, and colonnade (entrance). The facades of the two buildings are not the same, so similarities can be sought while reserving differences in the facade design of the newly built buildings. Opera script is an opera catalog and is a classic mark of Folk-art culture. We refined this fascinating process of

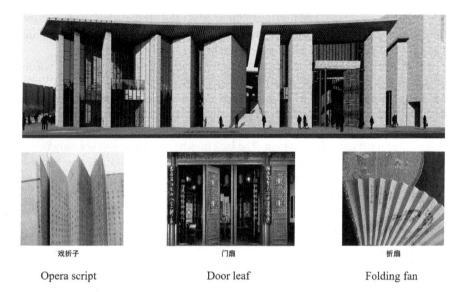

戏折子	门扇	折扇
Opera script	Door leaf	Folding fan

Fig. 4.6 Building facade language

folding and evolved it into a vertical plate language, making it look like a folding fan and a door leaf, with the meaning of hospitality. Under the eaves is a bronze-colored high-reflective metal grille, which is derived from the rafters under the roof of an ancient building, this is a simple way to respond to history (Fig. 4.6).

4.4 Conclusion

Inheriting the city's cultural heritage and continuing the historical memory of the city are indispensable parts of the city betterment and ecological restoration process.

Nowadays, China's urban development has moved from a large-scale new construction stage to a transformation and utilization stage of urban stock buildings. In the stage of large-scale new construction, although our urban environment and urban lifestyle have been greatly improved, and considerable economic benefits have been brought about, there is also a large amount of neglect or even destruction of history, humanities and urban original attributes. In China, we have seen thousands of cities that are the same, we have seen copy and paste, and many western city styles appear straightforwardly. On the surface, it brought the castles and European towns back to China, but it destroyed the natural and cultural environment. In the mechanized urban construction, it is regrettable that due to the cultural narrowness of the decision-maker, he imposes a personal aesthetic on the limited resources, such as cultural museums with very concrete and tacky colors. We should take the historical and human values of each piece of the land seriously, use reasonable design logic to respond to the urban memory of the site, let the building take root in that piece of land, and maximize the spatial value. If the city's memory cannot be continued, it will not only seal up the long history but also leave the city's life lacking the possibility of generating new vitality.

We can learn the big truth from the details of the whole group of buildings, which just like a miniature version of a vertical city and a composite mountain. While searching for memories in a hutong, people can inadvertently see the acrobatics being performed on the stage through the window between the brick walls. When they stop and watch, they can vaguely hear a burst of laughter and applause of the audience from the end of the hutong. The lack of urban memory will not only cover the history but also make urban life lack the possibility of generating new Energy. In this land with memories, we don't have to fill in the blank parts bluntly but, more carefully, fill in the blank parts softly and silently with vivid scenes.

References

Beijing Tianqiao Southern Acting Area. (2017). Guangzhou Zhujiang Foreign Investment Architectural Design Institute Co. Ltd. *Architectural Knowledge, 37*(15), 42–47.

Jie, H. (2018). Beijing Tianqiao Art Centre, Beijing, China. *World Architecture, 05*, 54–55. 116.

Jingqiu, L., & Shifen, L. (2020). Research on the protection and renewal of traditional houses in Luoyang Old City based on "Urban Double Repair" – Take the No.85 Courtyard of Luoyang Old Town as an Example. *Architecture & Culture, 08*, 104–105.

Ministry of Housing and Urban-Rural Development of the People's Republic of China. (2017). Guiding Opinions of the Ministry of Housing and Urban-Rural Development on Strengthening Ecological Restoration and Urban renovation. [EB/OL]. http://www.mohurd.gov.cn/wjfb/201703/t20170309_230930.html.2017.

Yucheng, T., & Xiang, J. (2020). Study on the renewal strategy of Xuzhou based on the background of urban double repair. *Chinese and Overseas Architecture, 9*, 91–93.

Zhongping, W. (2015). The acupuncture for urban fabric: Maximum effect through minimum intervention. *New Architecture, 03*, 4–8.

Part II
Sustainable Architecture Design and Technology

Chapter 5
The Suitability of Sustainable Design

Yehao Song, Yingnan Chu, Jingfen Sun, Xiaojuan Chen, and Dan Xie

5.1 Introduction

Sustainable design covers a wide range of subjects, involving environmental, economic, social, and many other issues. In contemporary China, the disequilibrium of development in different parts of the country caused a significant imbalance between the urban and rural areas. Grounded in the research of a sustainable design strategy that is suitable and localized for China, SUP Atelier developed a framework of suitability for sustainable architectural design (Fig. 5.1) including three focus points through practice (Song, 2018): firstly, the climate suitability closely associated with environment consciousness; secondly, the suitability of building and technology system closely associated with economic conditions; and thirdly, the cultural suitability closely associated with social equality. Projects by SUP Atelier designed and participated are based on the research of these three areas.

Y. Song (✉) · Y. Chu
Tsinghua University, Beijing, China
e-mail: ieohsong@tsinghua.edu.cn

J. Sun · X. Chen
Architectural Design and Research Institute of Tsinghua University, Beijing, China

D. Xie
Beijing Tongheng Urban Planning & Design Institute, Beijing, China

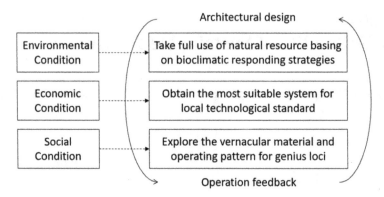

Fig. 5.1 Framework of suitability for sustainable architectural design

5.2 Climate Suitability

Climate suitability refers to the climatic responding design strategies based on the bioclimatic comfort of users in different climate zones. The concept of bioclimatic design has been established for nearly 60 years, through which time the research have made significant progress assisted by scientific development.

Within the category of bioclimatic responding strategies, those most relevant to architectural design are the consideration of sun, wind, water, and earth (Coch, 1998). These four elements generally covered the most important factors that can be interacted with during design process as well as initiating design strategies in environment related sustainable design. The use of natural resources in buildings – in terms of both energy and materials – has always been important within China's age-old architectural tradition (Hao & Song, 2016) and is also a key part of sustainable design in the country today. Under the premise of satisfying bioclimatic comfort needs in different climate zones and among different groups of people through architectural design, solar energy and natural ventilation are widely utilized to reduce the consumption of non-renewable energy, and sustainable natural materials are employed to reduce the use of artificial materials (Iyengar, 2015).

For example, during the summer months, the provision of shade is necessary, and it is possible to take advantages of the heat produced by direct sunlight at the same time. Utilizing the buoyance resulted from air being heated by the sun, a natural air circulation can be encouraged, therefore achieving passive cooling by letting out warm air inside the structures. During the winter months, the greenhouse effect provides passive heating for buildings, therefore reducing the heat load of the air conditioning system.

In China, natural air circulation has always been considered an important strategy for climate regulation: when the interior of a structure is over heated, a natural breeze could bring in the cooler air from the shaded area on the north side, therefore force out the heated and exhausted air. It not only regulates the thermal comfort, but also supplies a fresh air to the interior and reduces the energy consumption of air conditioning system. The two fundamental driving forces of natural air circulation – buoyance and wind pressure – are both directly related to architectural design (Hussain & Oosthuizen, 2012). Buoyance is commonly attained in architectural design either by means of tall wind tower that encourages air circulation through chimney effect or by means of Venturi Tube in the section of the building. Wind pressure can be created by adding architectural elements similar to the spoiler to buildings. The contribution of natural air circulation to energy saving and regulation of climate comfort can be achieved to a higher degree through the implementation of both.

The most important strategy to utilize water in the design process is the collection and use of rain water. The character of water that reduces temperature through evaporation creates a relatively comfortable micro-climate through which heat reduction can be realized. Among its many uses, rainwater can also be stored to provide source of water for plantations on the roof and surface of the building.

Soil is the most common material that contains embodied energy. It can be used for the basis of plantation growth, as well as in earth-sheltered architecture, which provide an efficient heat protection between the structure and its environment.

THE Studio is located in Gui'an Innovation Park, Guizhou province, southwest China. A collaboration between Qinghua University and BRE, this pilot technological demonstration project is two storey above the ground and 701 m^2 in area. The building consists of large exhibition space, conference area, and offices. It has been awarded BREEAM three star and is a LEED Platinum certified building. The layout of the building centers around the atrium, a spacious exhibition area, while the two wings contain relatively enclosed functional rooms.

The ridge of the roof accentuates further upwards to accommodate continuous skylights, not only enhances the upwards air circulation through chimney effect, but also provides a natural gentle lighting to the interior. The colored optical glass film brought a lively atmosphere to the atrium, cast a colorful iridescence shadow through glass and the wooden roof truss, changing through seasons and times.

The exterior of the building is a double-layered skin that responds to the climate (Fig. 5.2). The first layer is a simple glass wall with ventilation openings. The second layer consists of wicker woven surface of different density, designed base on the simulation result of sun radiation and wind pressure distribution. Four wicker woven units are arranged according to patterns and density in texture on the façade.

In the RSC building, completed in 2020, an overall energy-saving scheme has been implemented. The semi-open and open spaces, which serve the public, are air-conditioning-free, while only indoor areas, mostly used as offices with privacy requirements, are air-conditioned to minimize air-conditioning load. Sustainable goals are reached by the following two approaches.

On one hand, sun shading and ventilation design are specially made to adjust the microclimate. For example, none of the open areas, the semi-open full-height atrium, or the platforms are air-conditioned; instead, natural ventilation, daylighting, and sun-shading by the building itself have guaranteed the thermal comfort in these areas (Fig. 5.3).

Fig. 5.2 Envelop of THE Studio. (Credit: Xia Zhi)

Fig. 5.3 RSC building with semi-open atrium

On the other hand, a series of strategies are applied to air-conditioned areas to improve energy efficiency, including specially designed sun-shading systems and an alley-courtyard system which facilitates daylighting and ventilation. Functional areas like offices are air-conditioned yet considered for sun-shading and natural ventilation, with part of them earth-covered to make the most of the favorable conditions for vegetation and reduce energy consumption. Furthermore, the full use of natural light and ventilation, as well as innovative utilization of sun-shading parts, are implemented, showcasing our sustainability awareness that SUP has always maintained. Daylighting systems are implemented in each core area and earth-covered areas for ample natural light, and each courtyard can serve as a channel for ventilation, reducing air-conditioning load in spring and autumn with the help of sun-shading system.

5.3 Suitability of Building and Technology System

The suitability of building and technology system in architecture focuses on the feasibility of bioclimatic responding design strategies. In other words, it seeks to obtain the most suitable system for local technological standard and operation and maintenance ability, according to the economic development of the area. The challenge that faces sustainable design practice in contemporary China comes not only from the decision making of design and building technology; moreover, it comes from the difficulties of locating the most suitable building system specific to the area. Restricted by cost, level of complexity in construction, and standards of post completion operation and maintenance, the repertoire applicable to large urban area is impossible to replicate and implement in remote rural settlements, even though both areas are situated in the same climate zone and have the same bioclimatic requirements. In urban centers, sustainable design in China is in need of precedence that reflects high technical content, high cost, and high operational standard and maintenance, as well as having a bioclimatic comfort that is consistent to international standards (Zhang et al., 2020). In rural areas, precedence of low cost projects with low technical specifications is required. Additionally, a third type of precedence that represents the vast quantity of construction in contemporary China is in demand.

Longfor Sunda passive museum is an exhibition pavilion supported by Longfor Real Estate and Orient Sundar Windows Company, within which the sustainable architecture strategies and the edge-cutting building techniques could be introduced to the public. Meanwhile, the pavilion is targeted at the passive house certification. According to the PHI (Passive House Institute) standard, the building must be designed to have an annual heating demand as calculated with the Passivhaus Planning Package of not more than 15 kWh/(m2a); and the building must not leak more air than 0.6 times the house volume per hour ($n50 \leq 0.6 / h$) at 50 Pa (0.0073 psi) as tested by a blower door. Those strict standards make a lot of limitation of building insulation, area of openings and windows, air tightness of windows and doors, thermal bridges, and even the rational space form and shape coefficient.

In order to improve the heat performance, the north side of the building is sheltered with earth. The shape of the pavilion merges with the landscape topography, which make the pavilion's north facade completely vanished into the natural landscape. And the building's thermal losses are highly reduced due to earth shelter in comparison with ordinary building insulation. At the same time, the south side is made with curtain wall, which could work as solar collector in winter for passive heating and as a mirror reflecting clearly the surrounding environment to extend natural landscape (Fig. 5.4). In these ways, the north side of the building is hidden in the topographical landscape, and the south side of the building is hidden in the curtain walls' reflection of the trees and shrubs. The curtain wall, forming the whole south façade, could contribute to heat gaining in winter. While in summer, the sun-shading system functions automatically, tracing the orientation of the sunlight, preventing interior space from overheat. The skylight in atrium brings the sunlight in daytime and will be opened for the natural ventilation at night. The fresh air system also takes advantage of the particular shape of the interior space, matching the thermal press ventilation principle. The fresh air outlets are set in lower spaces, such as the corridor area in the north, and the bottom of the seating stairs in atrium and the return air inlet are set in the top of the south, the highest point in the interior space.

Also completed by SUP Atelier in 2015 was the Central Canteen of Tsinghua University, which houses not only a canteen for students and staff but also a career center for graduates. A corridor running from east to west across the building separates the canteen from the careers center. Seven oval-shaped skylights in the corridor bring natural air circulation through hot air pressure and wind pressure ventilation to the three-storey atrium in the warm months via openings designed on the side of each skylight. Abundant natural light and efficient energy savings are achieved alongside an effective cooling strategy in the summer (Fig. 5.5).

Tsinghua University southern campus canteen is a sustainably designed demonstration project located in the historical red brick buildings district of the campus. The west and north sides of the site are adjacent to the main routes of the campus, with students' dormitory located to its east and campus square to its south.

The building has a total area of 21,000 m², with 8000 m² above the ground in three story and three story of basements. The main function of the building serves

Fig. 5.4 Indoor environment of passive house museum. (Credit: Xia Zhi)

Fig. 5.5 Daylight distribution within the Central canteen of Tsinghua University. (Credit: Xia Zhi)

as student and staff canteen, as well as career center for graduates. A spatial corridor running from east to west across the building separates the canteen and the career center. Seven oval-shaped skylights in the corridor bring natural air circulation to the three-storey atrium in the warm months via ventilation openings designed on the side of each skylight. Abundant natural light and efficient energy saving are achieved alongside an effective cooling strategy in the summer. On the ground floor, an open and fluid route is designed to provide same level entrances to different interfaces spacially, creating a communicative public space for students and staff that are coming from different parts of the campus.

Indoor Playground of Yueyang County No. 3 Middle School is designed doubling as a lecture hall with the purpose of creating better sports space and invigorating the regularly planned campus. Yueyang is located in the hot-summer and cold-winter climate zone in China, where the weather is humid and rainy all year round. Natural ventilation and lighting serve as major sustainable strategies to improve thermal comfort and reduce the cost of equipment and maintenance. The integration design approach takes form, space, and sustainable strategies into consideration simultaneously (Fig. 5.6).

A skylight renders the ambience of the rostrum, and the space above the rostrum is heightened to accommodate a rainproof air-vent on the inclined façade without mechanical appliances. An array of operable doors at the bottom of southern and northern facades can boost natural ventilation, reduce humidity, and improve thermal comfort. A narrow alley is planned between the campus's southern wall and buildings along the wall. In summer, the wall's shade cools the air before it enters the building. Ample skylights on the jagged roof can provide enough natural light even in rainy weathers, while louvers on the jags take away the heat.

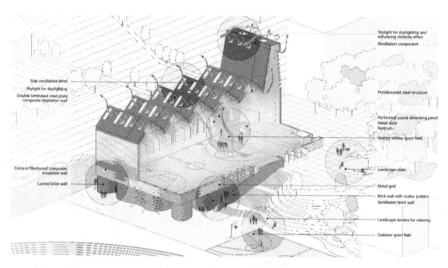

Fig. 5.6 Technique of Yueyang middle school indoor playground

Cast-in-situ methods and prefabrications have been both applied in the construction to alleviate the impact on the campus. At the lower part, the rubble stone retaining wall has been extended, while the construction was done through on-site construction with reinforced concrete and red bricks. At the upper part, the main space was built with prefabricated steelwork envelopes and roofs.

5.4   Embodying Cultural Diversity

Cultural suitability is another focus point stems from the unique condition of contemporary China. It refers to the most suitable materials, building technique and operating pattern for Genius Loci, and the contribution of the architecture toward the preservation of culture diversity (Song & Sun, 2018). In fact, the relatively imbalanced economic development in different areas of contemporary China, as well as the diversity in climate and geographic characteristics, provides many excellent possibilities and design precedence in sustainable design by way of vernacular architecture to achieve cultural diversity.

The Village Lounge in Shangcun, Jixi county, Anhui province, is another demonstration case for sustainable practice. The project collected and organized old black bricks, black tiles, stone and usable timber from the site to form landscape elements based on the original layout of the courtyard. The design solution is grounded on the principal of minimal intervention, adopted a layout of multiple units using common local material – bamboo – to construct 6 sheltered large space. Three in a row and 2 in a column, the six 5mx5m spaces form 3 sets of bamboo canopy with black awning, providing a shared space hosting the activities of the village locals and the visitors (Fig. 5.7).

Fig. 5.7 Village lounge of Shangcun, adopting bamboo as symbolic for local atmosphere

The project collected and organized old black bricks, black tiles, stone, and usable timber from the site to form landscape elements based on the original layout of the courtyard and infrastructure construction, such as the maintaining of the old MaTau walls, and the construction of the stone retaining walls with traditional techniques, meanwhile local craftsmen shared their ideas about the details, planting and decorations, which also made the lounge something that actually grew in the village. The bamboo umbrellas, apart from traditions, are built with modern architectural techniques by professional bamboo craftsmen, in order to ensure the durability of the bamboo components, and the other part of the lounge and the landscape is completely constructed by the villagers themselves.

The configuration of the bamboo umbrella and the arch of the black awning originated from an experiment to simplify the structure and reduce the scale of the roof. The black tiled vernacular buildings in the village have an average roof depth of 5–6 m. When viewing from the hill top, the canopy has a similar scale that blends into the fabric of the village, providing a public space for meetings and entertainment (Fig. 5.8).

The villagers' participation was essential. In the early stage of the project, the villagers cleaned up the site and precisely picked out old bricks, black tiles, stone plates, and timber, offering ample recycled materials for the projects. Then they watched the on-site work of the bamboo contractor and learned about the maintenance of bamboo, the utilization of modern tools, as well as the contractor's seriousness and carefulness during construction. All these inspired the local craftsmen to do their work with great enthusiasm and elaboration. As the project drew close to completion, many villagers have turned from onlookers to participants in assembling furniture, cleaning up the site, planting trees, and making decorations, truly becoming the owners of the lounge.

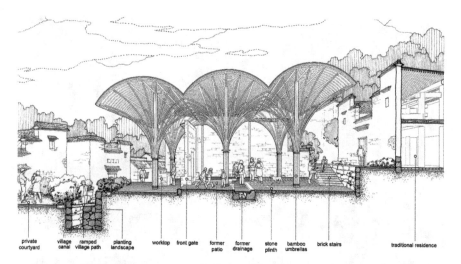

| private courtyard | village canal | ramped village path | planting landscape | worktop | front gate | former patio | former drainage | stone plinth | bamboo umbrellas | brick stairs | traditional residence |

Fig. 5.8 Implementation of local building technique of village lounge of Shangcun Village

At THE Studio in Gui'an, the climate-responsive double-skin facade system applies a local handicraft product: woven rattan. Although the outer layer of the skin, which consists of four densities of rattan units, is based on the results of simulations of solar radiation and wind pressure distribution on the building's surface, the rattan comes from local craftsmen, supporting the development of the ancient local rattan heritage. On the ground floor, an open and fluid route is designed to provide same-level entrances to different interfaces spatially, creating a communicative public space for students and staff who are coming from different parts of the campus. Landscaping outside the building continues the original site formation with respect to the existing plantation. The building sets back on the southeast corner in order to retain a historic Chinese parasol tree. The thoughtful approach to the history of the site helps to initiate a dynamic modern campus life.

The tea market in Shouning is another project exploring the combination of tradition cultural of construction with modern technique. The wooden arch bridge is a traditional bridge structure form in Fujian, China, which utilizes the principle of overlap between bamboo and wood to convert deflection into side thrust and solve the need for long spans. At present, the technique of wooden arch bridge has become China's intangible cultural heritage.

The roof secondary structure is made of bamboo, which is widely planted in Fujian and Zhejiang, China (Fig. 5.9). As a building material, due to its light weight and good toughness, the earthquake resistance of bamboo is very outstanding, and it is a widely accepted regional building material. Durable bamboo can be used in pavilions, observation decks, corridors, pontoons, etc. Secondly, the main structure of the building's large-span structure is inspired by the arched wooden bridge.

The external wall is made of rammed earth. The rammed earth is a common wall material of traditional houses in Fujian area, using stone foundation and compacted external wall. The color is mainly light yellow, and it has good durability and easy

Fig. 5.9 Tea market in Shouning under construction

operation. With the help of modern molds and instruments, rammed earth technology is currently simple and quick to operate.

The building is constructed with large quantity of sustainable recourses such as timber, steel, and straw panel. It also encourages the use of local materials and building techniques such as the application of traditional wicker waving on the skin of the building, and the utilization of local black stone on the interior floor and ashlar laying on the exterior fences. These strategies effectively reduce the carbon footprint of the building in its life cycle and create a unique vernacular architecture language.

5.5   Conclusion

Modern China has a complicated variety of climatic, economic and social condition, thus the sustainable design could hardly be simplified to a concise framework as a whole. The projects by SUP Atelier in this paper explore the interpretation of the climate suitable with natural resource, suitability of building technology, and cultural diversity enhance to be three main drivers of sustainable design, which demonstration the contemporary sustainable design strategies in China. Meanwhile, the process of "Design-Construct-monitor-feedback" is necessary for these sustainable strategies to be useful and appropriated, forming a loop of evidence-based design toolkit for the future practice.

Acknowledgments This work was funded by National Natural Science Foundation of China (Project No. 51678324, 52078264).

References

Coch, H. (1998). Bioclimatism in vernacular architecture. *Renewable Sustainable Energy Reviews, 02*(1–2), 67–87.

Hao, S., & Song, Y. (2016). The climate responsive mechanism of courtyard dwellings in hot-humid climate. *Eco-city and Green Building, 4*, 22–29.

Hussain, S., & Oosthuizen, P. (2012). Numerical study of buoyancy-driven natural ventilation in a simple three-storey atrium building. *International Journal of Sustainable Built Environment, 1*, 141–157.

Iyengar, K. (2015). *Sustainable architectural design: An overview*, Routledge.

Song, Y. (2018). Three Main drivers of sustainable design: A new architectural vernacular for China. *Architectural Design*, 74–79.

Song, Y., & Sun, J. (2018). Building Village lounge in bamboo towards a sustainable future at Shang Village a practice of participatory design. *Architectural Journal, 12*, 36–43.

Zhang, S., et al. (2020). Scenarios of energy reduction potential of zero energy building promotion in the Asia-Pacific region to year 2050. *Energy (Oxf)*, https://doi.org/10.1016/j.energy.2020.118792. Epub 2020 Sep 10. PMID: 32929299; PMCID: PMC7481845.

Chapter 6
Automatic Generation of 3D Building Models for Sustainable Development

Kenichi Sugihara

6.1 Introduction

A 3D city model, such as the one shown in Fig. 6.1 right, is important in urban planning and architectural design, e.g., BIM (Building Information Modeling). Urban planners may draw the maps for sustainable development. 3D city models based on these maps are quite effective in understanding what if this alternative plan is realized. To facilitate public involvement for sustainable development, 3D city models can be of great use. However, enormous time and labor has to be consumed to create these 3D models, using 3D modelling software such as 3ds Max or SketchUp. A complicated orthogonal polygon can be partitioned into a set of rectangles. The proposed integrated system partitions orthogonal building polygons into a set of rectangles and places rectangular roofs and box-shaped building bodies on these rectangles (Sugihara & Hayashi, 2008; Sugihara & Kikata, 2013).

Since technicians are drawing building polygons manually with digitizers, depending on aerial photos or satellite imagery as shown in Fig. 6.1 left, not all building polygons are precisely orthogonal. When placing a set of boxes as building bodies for forming the buildings, there may be gaps or overlaps between these boxes if building polygons are not strictly orthogonal. Our contribution is the new methodology for rectifying the shape of building polygons and constructing 3D building models without any gap and overlap (Sugihara & Murase, 2016; Sugihara & Murase, 2018).

In our proposal, after approximately orthogonal building polygons are partitioned and rectified into a set of mutually orthogonal rectangles, each rectangle knows which rectangle is adjacent to and which edge of the rectangle is adjacent to,

K. Sugihara (✉)
School of Information Media, Gifu Kyoritsu University, Ogaki, Japan
e-mail: mjsbp812@yahoo.co.jp

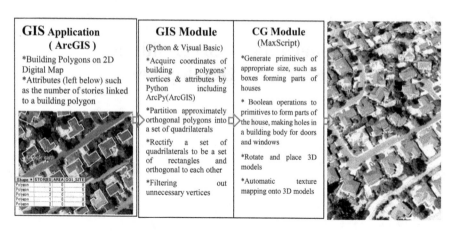

Fig. 6.1 Pipeline of automatic generation for 3D building models

which will avoid unwanted intersection of windows and doors when building bodies combined.

In this paper, in addition, for greater solar photovoltaic (PV) generation, a double leaned roof attached by PV arrays is automatically created. The sizes and positions, slopes of roof boards, and under roof constructions are clarified by designing the top view and side view of a double leaned roof.

6.2   Pipeline of Automatic Generation

As shown in Fig. 6.1, the proposed automatic building generation system consists of GIS application (ArcGIS, ESRI Inc.), GIS module and CG module. The source of a 3D city model is a digital residential map that contains building polygons linked with attributes data shown in Fig. 6.1 left, consisting of the number of story, the image code of roof, wall, and the type of roof (gable roof, hipped roof, gambrel roof, mansard roof, temple roof, and so forth). The maps are then preprocessed at the GIS module, and the CG module finally generates the 3D city model. As a GIS module, a Python program including ArcPy (ArcGIS) acquires coordinates of building polygons' vertices and attributes. Preprocessing at the GIS module includes the procedures as follows:

1. Filter out an unnecessary vertex whose internal angle is almost 180 degrees.
2. Partition or separate approximately orthogonal polygons into a set of quadrilaterals.
3. Generate inside contours by straight skeleton computation for placing doors, windows, fences and shop façades which are setback from the original building polygon.

4. Rectify a set of a quadrilateral ('quad' for short) to be a set of rectangles and orthogonal to each other.
5. Export the coordinates of polygons' vertices, the length, width and height of the partitioned rectangle, and attributes of buildings.

The CG module receives the preprocessed data that the GIS module exports, generating 3D building models. In GIS module, the system measures the length and inclination of the edges of the partitioned rectangle. The CG module generates a box of the length and width, measured in GIS module. In case of modelling a building with roofs, the CG module follows these steps:

1. Generate primitives of appropriate size, such as boxes, prisms or polyhedra that will form the various parts of the house.
2. Boolean operations applied to these primitives to form the shapes of parts of the house, e.g., making holes in a building body for doors and windows, making trapezoidal roof boards for a hipped roof and a temple roof.
3. Rotate parts of the house according to the inclination of the partitioned rectangle.
4. Place parts of the house.
5. Texture mapping onto these parts according to the attribute received.
6. Copy the second floor to form the third floor or more in case of building higher than 3 stories.

CG module has been developed using Maxscript that controls 3D CG software (3ds MAX, Autodesk Inc).

6.3 Polygon Partition Process

Figure 6.2 shows detailed process of polygon partition and shape rectification, generation of a 3D building model. When the vertices of a polygon are numbered in clockwise order, and the edges of a polygon are followed clockwise, an edge turns to the right or to the left by 90 degrees. It is possible to assume that an orthogonal polygon can be expressed as a set of its edges' turning direction; an edge turning to the "Right" or to the "Left."

A useful polygon expression (RL expression: edges' Right and Left turns expression) was proposed for specifying the shape pattern of an orthogonal polygon (Sugihara & Hayashi, 2008; Sugihara & Kikata, 2013). For example, an orthogonal polygon with 8 vertices shown in Fig. 6.2(a1) is expressed as a set of its edges' turning direction, RRLRLRRR where R and L mean a change of an edge's direction to the right and to the left, respectively.

The more vertices a polygon has, the more partitioning scheme a polygon has, since the interior angle of a "L" vertex is 270 degrees and two DLs (dividing lines) can be drawn from a L vertex. In our proposal (Sugihara & Hayashi, 2008), among many possible DLs, the DL that satisfies the following conditions is selected for partitioning.

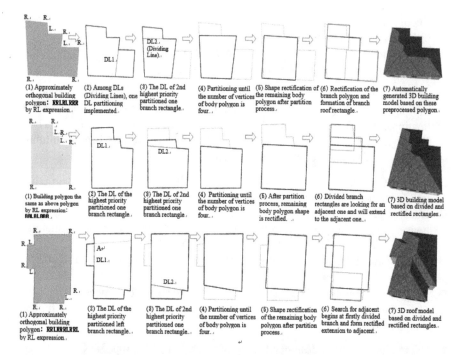

Fig. 6.2 Detailed process of polygon partition and shape rectification, generation of 3D model

1. A DL that cuts off "one rectangle".
2. A DL whose length is shorter than the width of a "main roof" that a "branch roof" is supposed to extend to, where a "branch roof" is a roof that is cut off by a DL and extends to a main roof.
3. A DL whose vertices are not shared by another DL.

The system executes the partitioning procedure as follows:

1. Classify the branches by the number of successive "R" vertices and the length of the edge especially incident to "L" vertex and dividing pattern.
2. Check whether the DL is satisfying three conditions or not, by measuring the distance between the DL and other edges right in front in the same polygon, which will be the width of the main roof.
3. If the DL satisfies the conditions and is given the highest priority, then the intersection position between the DL and edges is calculated.
4. Set the erase flags for the vertices of the branch that are removed from a body polygon, and a new vertex that is the intersection will be included by the body polygon.
5. Measure the edge length and inner angle of the polygon's vertices, and acquire edges' Right and Left turns expression (RL expression).

This partitioning procedure continues until the number of the vertices of the body polygon is four. In Fig. 6.2 (a2), the DL1 dividing a branch quad is a Forward

Dividing Line (FDL) in terms of polygon vertices numbering (clockwise). In Fig. 6.2 (a3) & (b3), a DL2 dividing a branch quad is a Backward Dividing Line (BDL) drawn in the opposite direction (counterclockwise). The dividing pattern is defined by this FDL or BDL or FDL and BDL.

The DL1 dividing a branch quad in Fig. 6.2 (c2) is given the highest priority for partitioning, since this partitioning reduces the vertex number of the body polygon by four. The DL1 dividing a branch quad in Fig. 6.2 (a2) and (b2) is given less priority for partitioning, as this partitioning reduces the vertex number by two. Thus, the system is giving each DL the degree of priority for partitioning, and the partitioning by the DL of the highest priority will be executed. Two DLs satisfying the first two conditions can be drawn from the vertex "A" shown in Fig. 6.2 (c2). But, the third condition excludes the longer DL, since the third condition demands a DL whose vertices are not shared by another DL, which will prevent unnecessary dividing. A shorter DL is selected for partitioning. After remaining body polygon's vertex number is four, the shape rectification begins by transforming the remaining polygon into a rectangle, as shown in Fig. 6.2 (a5) & (b5) & (c5). An active branch quad will start to search for an adjacent quad in the order from the lastly divided branch. The active quad will find a neighboring quad by using quads' vertices position of before rectification. Therefore, each quad instance has double vertices positions of before and after rectification. When forming the branch rectangle and branch roof rectangle, the system is using quads' vertices position of after rectification.

6.4   Shape Rectification

Specifically, the rectification procedure is implemented to the polygon shown in Fig. 6.3, which shows the process of polygon partition and shape rectification, automatic modelling. Before polygon partitioning, all edge length and edge inclination of the polygon are measured, and the length of all edges are sum up according to the snapped angle of all edge inclination. Then, the angle for a longest sum up edge length can be adopted as the "main angle" of the polygon, which will be then used as the inclination of all partitioned rectangles. After GIS module measuring the length and inclination of all edges of the partitioned polygon, i.e., a quadrilateral ("quad" for short), the edges are categorized into a long edge (w_L or edge12) and a short edge (w_S or edge23). A partitioned quad (quadrilateral) is numbered clockwise with the start point of a longest edge facing right as pt1 (a1, b1,..) or with the start point of a longest edge facing left as pt3 (a3, b3,..) as shown in Fig. 6.3a.

When a quad is cut off, the dividing pattern and by which edge of the quad is cut off, i.e., an "active edge" is saved at the quad. During the searching stage, an active quad will search for an adjacent quad by locating which quad the checking point on the active edge contains and then checking on which edge of the adjacent quad the checking point is included. In case of quad (1) in Fig. 6.3a, DL (a3a4) will be an active edge, and search for an adjacent quad. After searching and having found out the adjacent quad is quad (4) and the adjacent edge is m1m2 of quad (4), the mutual

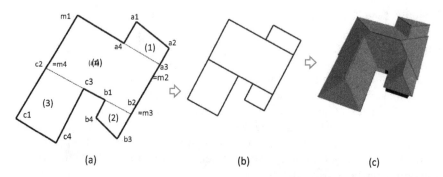

Fig. 6.3 Polygon partition and vertices numbering, shape rectification based on the generatrix, generation of 3D model. (**a**) Polygon partitioning. (**b**) Shape rectification. (**c**) Generated 3D model

vertex is a3(=m2), which the rectification procedure uses as a "standard position (generatrix)" for rectification, since this vertex is shared by two quads and could be an origin of local coordinates.

The rectified positions of the vertices of quad (1) are calculated as follows:

$$\mathbf{a1.x} = \mathbf{m2.x} + \mathbf{w\_S} * \cos\theta - \mathbf{w\_L} * \sin\theta$$
$$\mathbf{a1.y} = \mathbf{m2.y} + \mathbf{w\_S} * \sin\theta + \mathbf{w\_L} * \cos\theta$$
$$\mathbf{a2.x} = \mathbf{m2.x} + \mathbf{w\_S} * \cos\theta : \mathbf{a2.y} = \mathbf{m2.y} + \mathbf{w\_S} * \sin\theta$$
$$\mathbf{a4.x} = \mathbf{m2.x} - \mathbf{w\_L} * \sin\theta : \mathbf{a4.y} = \mathbf{m2.y} + \mathbf{w\_L} * \cos\theta$$

where θ is the main angle and w_S is the average length of two short sides of the quad and w_L is the average length of two long sides of the quad. In case of quad (3), the mutual vertex is c2(=m4), which the rectification procedure also uses as a standard position for rectification. The rectified positions of the vertices of quad (3) are calculated as follows:

$$\mathbf{c1.x} = \mathbf{m4.x} - \mathbf{w\_L} * \cos\theta : \mathbf{c1.y} = \mathbf{m4.y} - \mathbf{w\_L} * \sin\theta$$
$$\mathbf{c3.x} = \mathbf{m4.x} + \mathbf{w\_S} * \sin\theta : \mathbf{c3.y} = \mathbf{m4.y} - \mathbf{w\_S} * \cos\theta$$
$$\mathbf{c4.x} = \mathbf{m4.x} - \mathbf{w\_L} * \cos\theta + \mathbf{w\_S} * \sin\theta$$
$$\mathbf{c4.y} = \mathbf{m4.y} - \mathbf{w\_L} * \sin\theta - \mathbf{w\_S} * \cos\theta$$

The rectified positions of the vertices of a branch quad in other cases are calculated likewise according to the dividing pattern and "active edge."

6.5 Automatic Generation of a Double Leaned Roof House

Generated power of solar photovoltaic (PV) panels depends on the intensity of the sunlight and the installation condition of the panels such as panels' azimuth, pitch, surrounding environment, atmospheric temperature, and its location. The panel mounted large roof facing the sun at noon will have great solar PV energy. For example, "double leaned roof" houses shown in Fig. 6.6, which consists of two leaned roofs with upper dominant roof facing the sun at noon, are suitable for PV generation. The solar power generation of panels is proportional to the panel side perpendicular to the sunlight, one can know how much solar energy the buildings with panels can create by calculating the area of the perpendicular component of panels by the projection image. These houses are formed by placing two roof boards, under roof construction (prism), and a house body, depending on the top view (Fig. 6.5) and side view (Fig. 6.4). The placing of these parts of a building is implemented in following steps. After measuring the length and the inclination of the edges of the partitioned rectangle, the edges are categorized into a long edge (w_L) and a short edge (w_S). The vertices of the rectangle are numbered clockwise with the upper left vertex of a long edge being numbered "pt1" as shown in Fig. 6.3.

In a Constructive Solid Geometry (CSG) representation, we use volumetric primitives for the creation of 3D models. Each building part or primitive has its own control point ("cp") and local coordinates that control its position and direction. The position of a "cp" is different in each primitive. As shown in Fig. 6.5, for placing building parts properly, their "cps" are positioned at the point that divides edge12 and edge23 at an appropriate ratio. For example, a prism is used for the construction under roof boards. The "cp" of a prism lies in one of the vertex of the base triangle in an upright position when a prism is newly created.

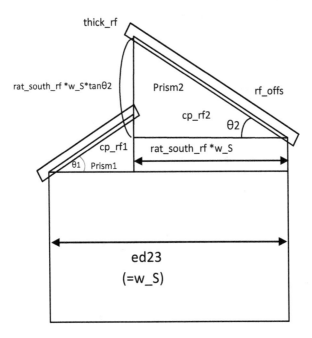

Fig. 6.4 Side view for double leaned roof; the position of the control point of two roof boards

Fig. 6.5 Floor plan for
double leaned roof; the
position of the control
point of two roof boards

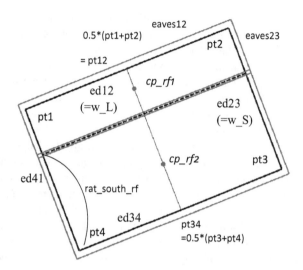

Fig. 6.6 Double leaned roof house automatically generated: houses for efficient solar photovoltaic generation

The top of a double leaned roof consists of two roof boards (two thin boxes). Since the "cp" of a box lies in a center of a base, it is placed on the point that divides the line through pt12 and pt34 at the ratio shown in ground plan (Fig. 6.5). The height of the "cps" of two roof boards is shown in the side view of a double leaned roof (Fig. 6.4).

To have larger south facing roof and get more solar power generation, the ratio of width of south facing roof (rat_south_rf) will be increased. The slope of two leaned roofs is given independently so that the system freely creates this type of roof, since the slope of the roof is also important factor for solar power generation. The width and slope of two leaned roofs will decide the height of top line of these roofs. If the difference in height between these top lines is greater than a certain length, then the prism (under roof construction) will be in Boolean operation to have holes for windows, and windows will be installed between two leaned roofs as shown in Fig. 6.6, which is the town block full of double leaned roof houses automatically generated.

6.6   Conclusion

To facilitate public involvement for sustainable development, 3D models simulating current or future cities by a 3D CG can be of great use. 3D city models are important in environmentally friendly urban planning that will use solar photovoltaic (PV) generation. However, enormous time and labor has to be consumed to create these 3D models, using 3D modelling software such as 3ds Max or SketchUp.

In order to automate laborious steps, we proposed a GIS and CG integrated system that automatically generates 3D building models, based on building polygons or building footprints on digital maps, which show most building polygons' edges meet at right angles (orthogonal polygon). A complicated orthogonal polygon can be partitioned into a set of rectangles. The proposed integrated system partitions orthogonal building polygons into a set of rectangles and places rectangular roofs and box-shaped building bodies on these rectangles.

In this paper, for greater PV generation, we clarify the structure of double leaned roofs attached by PV arrays, which is made up of two leaned roof boards and prisms (under roof construction). The sizes and positions, slopes of roof boards, and main under roof constructions are made clear by designing the top view and side view of a double leaned roof house.

References

Sugihara, K., & Hayashi, Y. (2008). Automatic generation of 3-D building models with multiple roofs. *Tsinghua Science & Technology, 13*, 368–374.

Sugihara, K., & Kikata, J. (2013). Automatic generation of 3D building models from complicated building polygons. *Journal of Computing in Civil Engineering ASCE (American Society of Civil Engineers), 27*(5), 476–488.

Sugihara, K., & Murase, T. (2018). Building polygon rectification for automated 3D building models. *IEEE International Conference on Signal Processing Proceedings (ICSP)*, pp. 1065, https://doi.org/10.1109/ICSP.2018.8652494.

Sugihara, K., & Murase, T. (2016). Shape rectification of building contour for automatic generation of 3D building models. *Journal of Japan Society of Civil Engineers, Ser. F3 (Civil Engineering Informatics), 72*(2), I_167–I_174. https://doi.org/10.2208/jscejcei.72.I_167

Chapter 7
Performance-Driven Digital Design Methods of Early-Stage Architectural Design: Investigations on Auditoriums and Room Acoustics

Shuai Lu

7.1 Introduction

With the severe environmental and energy crises and people's growing demands for comfortable and healthy living conditions, building performance, such as energy consumption, daylighting, ventilation, and room acoustics, has become an increasingly crucial issue for architectural design. As design decisions made in the early design stage not only have a high impact on aesthetic and programmatic concerns of buildings, but are also of paramount importance to building energy efficiency (Attia et al., 2012; Markku, 1998), architects should take performance issues into account from the starting point of architecture design. Rules of thumb and other empirical knowledge could be sufficient for simple design problems, but for more complicated problems architects may well need additional aid to make wise design decisions on early-stage design tasks such as orientation, layout, massing, fenestration, and shading (Tian et al., 2018). To better support decision-making and guide the design toward high energy efficiency, various design aiding methods and tools have been intensively investigated in precedent research, including rapid predictions of building energy efficiency indicators, design optimization driven by building performance, sensitivity analysis revealing the influence of design parameters on building energy efficiency, and advice on design moves to pursue ideal building performance (Lin & Gerber, 2014). Unlike traditional building performance simulation that is more frequently used in late design stages by technical consultants, these

S. Lu (✉)
School of Architecture, Design and Planning, the University of Sydney, Darlington, NSW, Australia

School of Architecture and Urban Planning, Shenzhen University, Shenzhen, People's Republic of China
e-mail: shuai.lu@sydney.edu.au

methods with various cutting-edge technologies integrated are specifically desig-
nated for architects to improve the energy efficiency of their design and address
common challenges in the early design stage such as repeated changes of design,
vast design space, lack of information, and time-consuming performance analysis
process (Østergård et al., 2016).

Among various aspects of building performance and different types of buildings,
this research focuses on the acoustic performance of auditoriums. An auditorium is
a container of performing arts, and frequently it also serves as a landmark of a city
or even of a state. A perfect acoustic environment for performance is the key to a
successful auditorium design. However, in most cases, architects are not proficient
in acoustic issues and thus need assistance from acousticians during early-stage
auditorium design. The cooperation between architects and acousticians is likely to
cause delays in design progress and discontinue the design thinking of architects, as
different groups of people are involved and information needs to be transferred
between them (Apfel, 1992; Beranek, 2001). As a result, the efficiency of architects
is likely to be affected, and the possibility of achieving novel and acoustically
appreciated designs in a limited time is reduced (More details in Sect. 7.2). It would
be undoubtedly beneficial for architects if they could have a general idea of the
acoustic quality and possible improvement options of their design before they talk
to acousticians, to reduce the efforts spent in repeated communication, and have
more time in design refinement.

To fulfill architects' demands of acoustic assistance, a new performance-driven
digital design process of auditoriums is proposed and implemented here, aiming to
provide architects with real-time architectural and acoustic feedback in the early
design stage. The proposed design process is composed of three parts:

1. A parametric model of auditoriums in Rhinoceros, which can generate various
   auditorium designs automatically (architectural feedback).
2. An interface connecting Rhinoceros and CATT, which supplements the input,
   manipulation, and output of CATT to facilitate architects' utilization (acoustic
   evaluations).
3. An acoustic aiding tool based on Support Vector Machine, which can provide
   quick acoustic tests, the acoustic potential of forms, and suggestions for design
   modifications (acoustic feedback).

7.2 Proposed Design Process

The current workflow of early-stage auditorium design could be simplified as a
continuous trial-and-error loop of analysis, synthesis, and evaluation with the
involvement of architects and acousticians according to the Bruce Arche Model
(Lawson, 2005) (Fig. 7.1). Architects need acousticians to test the design they propose and provide suggestions for further modifications if necessary, as auditoriums
have critical requirements of acoustics. However, (1) this feedback loop can hardly

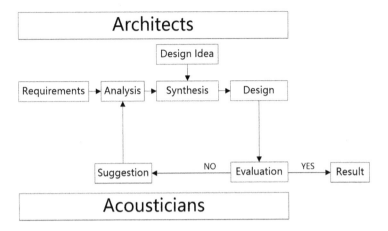

Fig. 7.1 Current process of auditorium early-stage design (by the author)

be timely in reality, because different groups of people are involved and a lot of information needs to be transferred between them. This not only affects the continuity of architects' design thinking but also takes up the time that architects could spend on refining the design or developing neoteric design ideas. (2) Architects and acousticians have different ways of thinking and value judgments. As Leo Beranek indicated, architects want auditoriums to be unique and inspiring, while acousticians tend to learn from highly successful halls (Beranek, 2001), thus conflicts may occur. In some cases, architects are the dominating side, then the design results could have some acoustic defects that are difficult to be fully remedied by acousticians in the late design stage when major aspects of architectural design are completed; in other cases, where acousticians are the leading side, the design results often share similar features due to preferences of the acousticians, while creative ideas of architects could be abandoned regretfully. Therefore, improvements in the current design process are needed.

To remedy the defects of the current design process, this research aims to raise acoustic aiding methods that provide real-time acoustic evaluations and design suggestions for architects. Taking this aiding method into account, theoretically, the design process of auditoriums could be improved to a three-party loop: architects, acousticians, and computers (Fig. 7.2). Initially, architects can work with computers, which will make quick acoustic evaluations and modification suggestions on the design they propose. Even though not perfectly precise, this quick evaluation and modification suggestions can help architects make reasonable design progress: if the acoustic evaluation is undesirable, architects can immediately start to rethink their design idea with the help of modification suggestions; otherwise, if the acoustic evaluation is desirable, acousticians can take over the work and verify the acoustic quality of the design more carefully, since there might be some issues that cannot be fully recognized by computers. If the acousticians think there is still some potential to further improve acoustic performance, they can provide architects with specific suggestions for further design modifications.

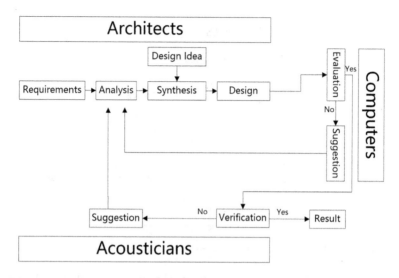

Fig. 7.2 Proposed process of auditorium early-stage design (by the author)

In this new design process, architects can avoid frequent delays and possible conflicts in communication with acousticians, and have more time on refining the design or developing novel design ideas. Thus, outstanding designs are more likely to be achieved in limited design time. For acousticians, it should be emphasized that their importance is not reduced at all. Now they are released from constantly testing different designs superficially, and can focus on pre-selected ones by computers to make more detailed examinations and provide more specific suggestions. In other words, the aim of the new design process is by no means to "substitute" acousticians, but to make the work easier and more efficient for both architects and acousticians.

7.3 Implementation of Architectural Feedback

The architectural feedback is provided by a parametric model of auditoriums. The idea of parametric models is to model buildings by defining essential rules instead of actual geometries. It can automatically provide a family of designs that meet requirements quickly, saving architects' efforts in repeatedly modifying geometrical models to get different design options.

Here the parametric model is built by "Component-Based" Method which is a method that includes different types of building components (walls, doors, stage, etc.) and their inter-connections. Based on this method, a parametric model of shoebox concert halls is implemented in Rhinoceros (V.5 SR11) with Python Editor (Fig. 7.3). The result shows that a large variety of designs can be generated. Any concert hall with vertical walls and no curvy components can be generated. Meanwhile, architects' ideas can be turned into designs simply by drawing drafts or

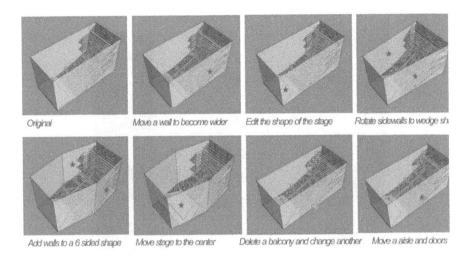

Original	Move a wall to become wider
Edit the shape of the stage	Rotate sidewalls to wedge sh.
Add walls to a 6 sided shape	Move stage to the center
Delete a balcony and change another	Move a aisle and doors

Fig. 7.3 Designs that can be generated by the parametric model (star indicates changed component) (by the author)

modifying components. Every component can be modified directly, accordant with architects' design customs. Therefore, this parametric model is effective and could be used in design practice.

7.4 Implementation of Acoustic Evaluation

The method to implement acoustic evaluation is to build an interface bridging Rhino and CATT, which improves architects' utilization of acoustic simulation in the input, manipulation, and output.

For input of simulation, this interface can facilitate information collected automatically. It can convert all geometries of Rhino into a .txt file following CATT rules. It automatically selects 2 sources on the stage, and 1 receiver in every 6 rows and 12 columns all over the audience area (Fig. 7.4). It provides two possibilities to choose the material, the first one is to choose from a material list for each kind of components; the second is to use the materials of an existing successful auditorium (Fig. 7.5).

For manipulation, the interface automates and standardizes the process of simulation. It automatically manipulates CATT in the background using scripting. Two simulations are conducted for one design with different sources. Simulation parameters are fixed as recommended by CATT manual for a quick test (10000rays, 2000 ms echogram).

For output, the interface simplifies and visualizes the results. For concert halls, RT, EDT, BQI, G, ITDG, and C80 are selected as acoustic indices according to the recommendations of Leo Beranek's research (LF is not used for its relevance to

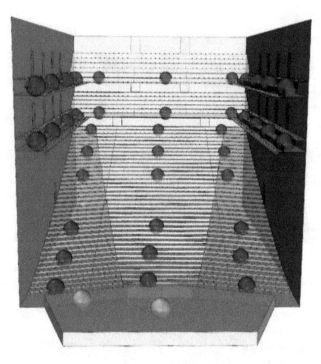

Fig. 7.4 Distribution of the sources (green) and receivers (red) (by the author)

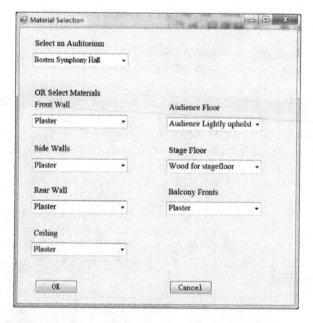

Fig. 7.5 User interface to input material information (select materials for each kind of components or select an auditorium) (by the author)

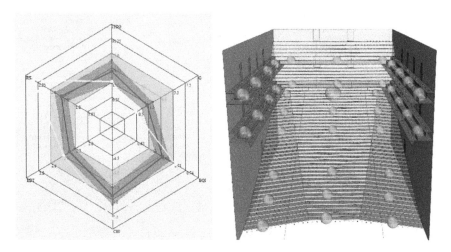

Fig. 7.6 The radar chart and the distribution map that show architects the simulation results (by the author)

BQI. BR and SDI are not used for their not relating to form design but material design). The recommended ranges of the indices are also from Leo Beranek's research. The results are visualized by two graphs, a radar chart and a distribution map. The radar chart shows the mean value, standard deviation, and the acoustically "worst point" of the design. And the distribution map shows the acoustic qualities of different receivers, and the position of the worst point (Fig. 7.6).

7.5 Implementation of Acoustic Feedback

The goal of the acoustic feedback part is to providing design suggestions based on acoustics, which is implemented by a two-stage methodology that integrates parametric model, acoustic simulation, and Support Vector Machine (Fig. 7.7). Stage 1 aims to acquire prediction models that can calculate the acoustic parameters of a certain design precisely and quickly. Auditorium designs (including shape and material information) are generated by the parametric model introduced in Sect. 7.3. Designs are then simulated by CATT Acoustics through the interface presented in Sect. 7.4, with a set of averaged room acoustic parameters derived. Based on the simulation results, prediction models of acoustic parameters based on design parameters are established using the Support Vector Machine algorithm. Stage 2 implements the functions of the aiding methods based on the prediction models. To generate design suggestions, the parameters of the current shape are changed in a certain pattern determined by the form of design suggestions to generate a series of new shapes, and the function of acoustic evaluation of auditorium shapes is applied to the new shapes. Then design suggestions can be derived from analysis of the results.

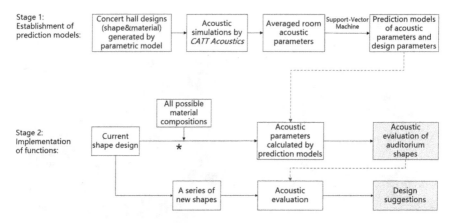

Fig. 7.7 Methodology of the acoustic feedback part (by the author)

Following the methodology above, the proposed functions of acoustic design suggestions for concert hall early-stage design are implemented in Rhinoceros (V.5 SR11) with its C# (for massive calculation) and Iron Python (for interface development) development kits. The following functions are included:

7.5.1 Variation Tendency of Acoustic Parameters

This function indicates the variation tendency of acoustic parameters with the change of certain shape parameters so that architects can understand that which shape parameters can be adjusted if certain acoustic parameters need to be improved. For a given shape, (1) its shape parameters for analysis are incremented and decremented for a small step (2% of the length of the parameter range by default, and can be modified) several times successively (20 times for both increment and decrement by default, and can be modified). (2) The acoustic parameters of all the modified shapes are calculated by the prediction models acquired in Stage 1 of the implementation methodology. (3) The variation tendency of the acoustic parameters against shape parameters is visualized.

An example chart that indicates the variation tendency of the acoustic parameters against selected shape parameters is shown in Fig. 7.8. Each chart represents one acoustic parameter, and each line color represents a shape parameter. The X-axis indicates the variation ratio of the shape parameters, while the Y-axis shows the value of the acoustic parameter. The red dots indicate the acoustic parameters of the current design, while the green areas show the acceptable ranges of acoustic parameters. With the help of these charts, architects can get ideas about which shape parameters to modify and how to modify, to fix the acoustic parameters that are currently undesirable.

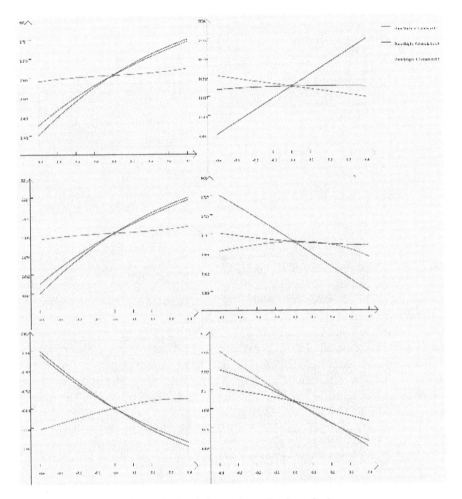

Fig. 7.8 An example of the result of variation tendency (by the author)

7.5.2 Recommendation of Similar Shapes

This function recommends a series of shapes that are similar to the current design but have desirable acoustics for architects so that architects can improve the acoustics of their current design directly by choosing from options of similar shapes. For a given shape, (1) each of its modifiable parameters is incremented and decremented for a certain step (5% of the length of the parameter range by default as 2% step could be too small to give distinguishable shapes, and can be modified) several times successively (5 times for both increment and decrement by default, and can be modified) so that its possible values are acquired. (2) All possible combinations of shape parameters (i.e., all possible shapes after modification) are enumerated, and

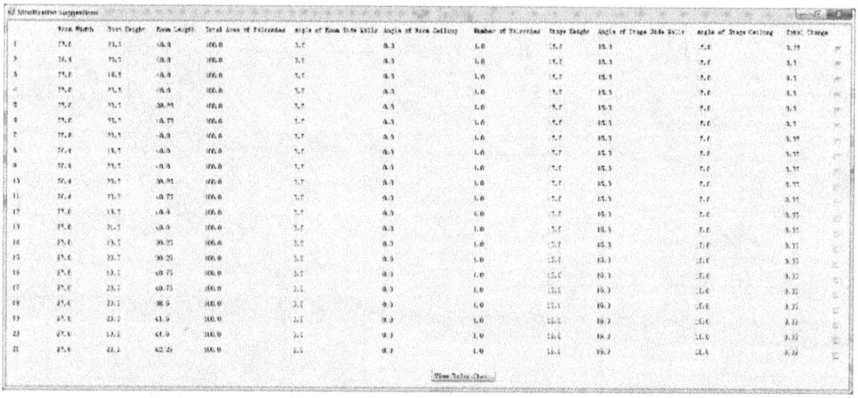

Fig. 7.9 An example of the result of similar shape recommendation (by the author)

all possible shapes are acoustically evaluated by the prediction models acquired in Stage 1 of the implementation methodology to determine if they are acoustically desirable. (3) The acoustically desirable shapes are ranked according to their similarity with the current shape. For each shape, the deviation of each shape parameter from the current shape is calculated. Then the ratio of this deviation to the range length of the corresponding shape parameter is calculated. Finally, the shape with a smaller sum of the deviation ratios is ranked in the front (as they are more similar to the current shape), while the shape with a larger sum of the deviation ratios are ranked in the back (as they are less similar to the current shape). (4) The ranked acoustically desirable shapes are displayed as a list.

An example list of ranked acoustically desirable shapes is shown in Fig. 7.9. Architects can improve the acoustics of the current design directly by choosing from options of similar shapes. In addition, architects can also check the box of the shape that they are interested in and click "View Radar Chart." Then the detailed acoustic performance of the shape will be shown (similar to the radar chart in Fig. 7.6).

7.5.3 Local Optimization

During shape creation, architects may have a vague and intuitive idea of the shape they want, or in other words, a range rather than a specific value for certain shape parameters. At this circumstance, acoustics can be a driven factor to determine the undecided shape parameters, so that the acoustics of the shape can be optimal on the premise that architects' design inspiration is maintained. The function of "local optimization" implements this idea and reveals the shape with optimal acoustics within architects' predefined ranges of shape parameters: (1) For each undecided

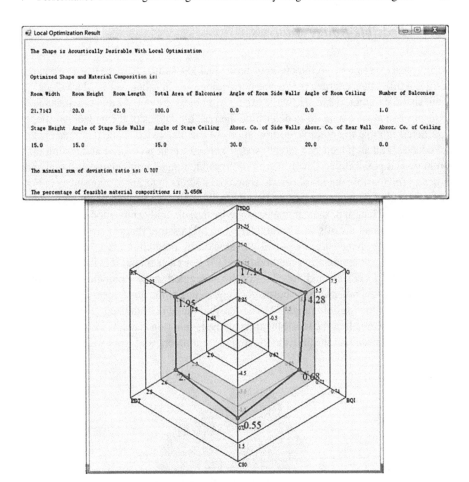

Fig. 7.10 An example of the result of local optimization (by the author)

shape parameter, its possible values are first enumerated within the predefined range with a small gap (2% of the length of the parameter range by default, and can be modified). (2) All possible combinations of shape parameters (i.e., all possible shapes within the predefined ranges) are enumerated, and all possible shapes are acoustically evaluated by the prediction models acquired in Stage 1 of the implementation methodology. (3) The shape with the best acoustics is searched and displayed for architects.

An example result of the local optimization function is displayed in Fig. 7.10. With its help, architects can turn their vague idea into an exact shape design with an ideal acoustic performance.

7.5.4 Feasible Range Analysis

During shape creation, architects may have decided some characteristics of shape design, while other characteristics are not decided, or in other words, the values of some shape parameters are set, while other parameters are not. At this circumstance, acoustics can also be a factor to determine the feasible ranges of the undecided shape parameters, so that the possible combinations of the undecided shape parameters can be reduced, and architects can largely spare their efforts in proposing and testing different design possibilities. The function of "feasible range" implements this idea and indicates the feasible ranges of certain shape parameters in terms of acoustics: (1) For each undecided shape parameter, its possible values are first enumerated with a small gap (2% of the length of the parameter range by default, and can be modified). (2) All possible combinations of shape parameters (i.e., all possible shapes) are enumerated, and all possible shapes are acoustically evaluated by prediction models acquired in Stage 1 of the implementation methodology. (3) For each shape, if it is acoustically desirable, then its corresponding values of the undecided shape parameters are within the feasible range; otherwise, its corresponding values of the undecided shape parameters are out of the feasible range. (4) The feasible range is visualized.

An example result of the feasible range analysis function is displayed in Fig. 7.11. Each axis represents an undecided shape parameter, and the shadowed area

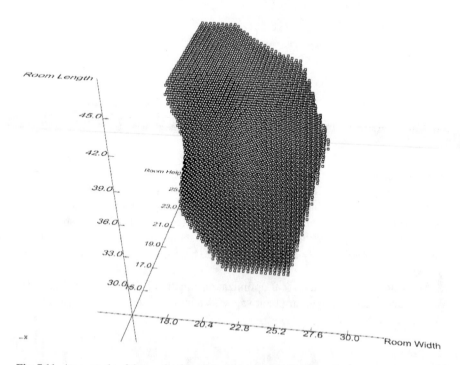

Fig. 7.11 An example of the result of feasible range analysis (by the author)

represents the feasible ranges in terms of acoustics. It should be noted that due to the limitation of visualization methods, only the feasible ranges of 1, 2, or 3 shape parameters can be analyzed by the current function.

7.6 Discussion

In this work, the definition of acoustically desirable shapes is based on six objective parameters that are widely used and have well-recognized recommended ranges. These quantitative parameters are effective to reflect major aspects of the acoustic performance of concert halls. Moreover, they can be used by architects to judge if the acoustic performance is desirable easily (by comparing the actual values and recommended ranges) and are not likely to be misleading. However, it has to be admitted that there are still some acoustic features that cannot be captured by these parameters. For example, the temporal envelope of early reflections and the distribution uniformity of sound energy are also important in concert halls, but there are no reliable quantitative parameters or recommended ranges to reflect these characteristics effectively. If these characteristics are included in the definition of acoustically desirable shapes, they could make it difficult for architects to make judgments that whether a shape is acoustically desirable, or could mislead architects to abandon acoustically acceptable designs. There is also some existing research trying to quantify these acoustic characteristics, for example, using the autocorrelation function to represent the distribution uniformity of sound energy. However, these methods have not been widely recognized yet and provided no reliable desirable ranges. As a result, the acoustic characteristics that are unable to be covered by existing room acoustic parameters are decided to be excluded in this research.

This limitation of the definition of acoustically desirable shapes could cause some shapes that are acceptable from the viewpoint of acoustic parameters but fail to meet other acoustic characteristics to be judged as acoustically desirable. However, these unsatisfactory designs can be recognized by acousticians, since every design approved by architects will be further verified by acousticians as indicated in Sect. 7.2. Therefore, this situation is obviously better than misleading architects to abandon promising designs. Moreover, even if new acoustic indices are added in the acoustic evaluation process, it is still impossible to quantify and include all acoustic characteristics of auditoriums, as room acoustics is still regarded as a mixture of art and science at the current stage, so verifications of acousticians are still indispensable.

In short, although the acoustic aiding method proposed and implemented in this research can help architects get a general idea of the acoustic quality and possible revision methods of their current design, and thus improve their efficiency and save their time spent in frequent communications, the aiding method by no means can "replace" acousticians, as limitations of its results are inevitable. All designs that are screened by the aiding method must be further examined by acousticians in detail.

7.7 Conclusion

Herein, a new auditorium design process involving parametric models, acoustic simulation, and machine learning is proposed and implemented, aiming to provide real-time architectural and acoustic feedback for architects. Several conclusions can be drawn from this research.

1. Component-Based Method is an efficient approach to develop parametric models of auditoriums, which can reliably provide architects with more varied and valid designs than existing methods, thus fulfilling the demand for architectural feedback. The feedback delays can be as short as 10 sec when based on Rhinoceros using a common computer.
2. A combination of integrated acoustic simulation and machine learning models is capable to provide acoustic feedback for architects. Acoustic evaluation and different ways of acoustic-based design suggestions (including variation tendency of acoustic parameters, the recommendation of similar shapes, local optimization, and feasible range analysis) can be implemented.
3. The acoustic aiding methods facilitate the improvement of the process of early-stage auditorium design. Architects' time and efforts in attempting infeasible design ideas and frequently communicating with acousticians can be largely saved, and consequently creative and acoustically desirable designs are more likely to be achieved in limited design time. However, the aiding method cannot replace acousticians and their detailed acoustic examinations are indispensable at all times.
4. More acoustic criteria should be included in future research (more acoustic parameters, impulse response, etc.), and verifications of the effectiveness in actual design practice are also needed.

References

Apfel, R. E. (1992). *Deaf architects & blind acousticians: Challenges to sound design*. Madison, CT.
Attia, S., Gratia, E., De Herde, A., & Hensen, J. L. M. (2012). Simulation-based decision support tool for early stages of zero-energy building design. *Energy and Buildings, 49*, 2–15.
Beranek, L. (2001). Concert hall acoustics. *Architectural Science Review, 54*(1), 5–14.
Lawson, B. (2005). *How designers think: The design process* (5th ed.). Architectural Press.
Lin, S. H., & Gerber, D. J. (2014). Evolutionary energy performance feedback for design: Multidisciplinary design optimization and performance boundaries for design decision support. *Energy and Buildings, 84*, 426–441.
Markku, J. (1998). IEA-BCS ANNEX 30, bringing simulation to application, subtask 2: Design process analysis, final report. FaberMaunsell Ltd on behalf of the International Energy Agency.
Østergård, T., Jensen, R. L., & Maagaard, S. E. (2016). Building simulations supporting decision making in early design – A review. *Renewable and Sustainable Energy Reviews, 61*, 187–201.
Tian, Z., Zhang, X., Jin, X., Zhou, X., Si, B., & Shi, X. (2018). Towards adoption of building energy simulation and optimization for passive building design: A survey and a review. *Energy and Buildings, 158*, 1306–1316.

Part III
Health and Human Considerations

Chapter 8
Vital Signs Revisited in the Tropics: Through the nus-cdl Tropical Technologies Laboratory

Stephen Siu-Yu Lau, Abel Tablada, Siu Kit Lau, and Chao Yuan

8.1 Introduction

Charles (Cris) Benton and other teachers of architecture initiated the Vital Signs project (Benton & Kwok, 1995; Kwok et al., 1997) as a way to connect designers with the Earth in the procurement of buildings for human activities. At the university where they taught, they shown students ways to connect and interact with nature, by listening, breathing, seeing, and experiencing those external changes that over time, have shaped and affected the properties of, changed the operation modes of building enclosures and their support systems. Many of these external changes took place in a flux of seconds or in the form of cyclic changes that induced phenomenal changes of the physical environment we live in. As we looked back, such works by educators are precursor to a unique branch of subject known as architectural science which is expressively created for meditating between human users and the environment, later translated into a new science known as Post-Occupancy Evaluation POE or Building Use Survey BUS (Turpin-Brooks & Viccars, 2006). Rightly, Benton is one of these pioneers who reminded the designer the practice of environmental design by research as a scientific way to reaching out to nature – by

---

The Bartlett School of Architecture changed its name to School of Environmental Studies when Llewelyn Davies was the Professor of Architecture (1960–69). He reformed the architectural teaching curriculum from Beaux-Arts approach to the Bauhaus approach and introduced science, multidisciplinary research. He raised major donation in the design and construction of a new building known as the Wates House. The Wates House was instrumental by the introduction of a basement floor devoted for scientific laboratories (White, 1826)

S. S.-Y. Lau (✉) · A. Tablada · S. K. Lau · C. Yuan
Faculty of Architecture, Technological University of Havana J.A.Echeverria, Havana, Cuba
e-mail: ssylau@hku.hk

observing, capturing, recording, and measuring often minutes changes. In essence, the vital signs are a replica and tribute for our ancestors who practiced similar acts of synchronizing with the changing universe, through changes of weather and climates which had direct bearings on the food production and safety for human settlement, since thousands of years ago. In the contemporary context, practicing with similar signs helped city and building designers apprehend how changes occurred and emerged in the first instance and learning how to predict their threats on human settlement as a result. Architectural science adds to the accumulation and formation of a body of knowledge made possible by science, physics, psychology, and other knowledge dealing with "climate, buildings, and man" in essence. In today's words, such knowledge makes possible the delivery of wellness for the human user of building, under the auspices of care for hygiene, health, restoration, and relief of stress and strength, i.e., the state of fitness of our body and mind. In the early 1990s marked the beginning of post occupancy studies in the education sector in the North America. On the other hand, the indulgence on building physics and related science as it was described started since the post WWII periods in Europe by research institutions and academia, notably the Building Research Station (today, the Building Research Establishment BRE, U.K. (2020)). On the academic side, the University College London was pioneer when it changed its name from School of Architecture to that of School of Environmental Studies under the professorship of Llewelyn Davies who introduced multidisciplinary research and social science to the architecture curriculum. Initially known as environmental or energy conscious design, particular attention focused on the interactions between climate, weather, building, material, and construction, i.e., the physiological and psychological responses of human user under different mode of activities, typically begun by the study on dwellings, hospitals, and workplaces. In contemporary world, these concerns have evolved into a set of criteria, resulting in commercially developed sets of guidelines for ensuring building and user performance under such tradenames as Wellness (McArthur & Powell, 2020) or national standard under the topics of Healthy Building (International WELL Building Institute, 2020).

POE marked the systematic studies on using user feedbacks as a premise to evaluate building performance for the sake of informing the designer at the initial stages of building design. In the preface of the report written by Lynda Stanley, Director of the Federal Facilities Council, "Learning from our buildings: A state-of-the-practice of post occupancy evaluation, Federal Facilities Council Technical Report No.145 (2001)," she gave the background for the POE study which was initiated by the federal building owner as early as the mid-1980s (1986). Since then, POE study has continued and gradually gained recognition by the design profession as a useful feedback pathway. Thirty years or so later, POE researchers have expanded on the significance of POE as a design tool by combining the dual concepts of "occupancy" with "building performance" evaluation (Preiser et al., 2014), while others have expanded its coverage from Pre- to Post-Occupancy Evaluation (Preiser & Schramm, 1997).

In 2016, the National University of Singapore announced a plan to design and construct a campus-based field laboratory for investigating climatic responsive architecture. Besides acting as a test-bed, it is designed to be a digitalized, environmental data center of the Tropics for the educational and practical ideologies discussed in the preceding paragraph. Given its location on the Kent Ridge Campus, it is adjacent to an exclusively built-up of tall, dense, public, and private housings being neighbor to the university staff quarter on which the laboratory is located. In essence, this setting mimics that of an urban environment typical of the Southeast Asian city norm featuring tallness, compactness, and mix land use. The laboratory has been designed by scaling down the otherwise full-sized mock-up of public housing. Designed to be a square plan incorporating four facades, it encompasses a total of 12 chambers (8 internal and 4 external) resembling the typical public housing building facade with fenestration and a similar facade but with a recessed balcony. On each of the fenestrations, different projections of sun-shading devices are incorporated with the external facing windows to provide sun protection. Almost all of the fenestrations are thus home to specially designed arrays of external plotted planters for the growing of edible plants of various species. The planters contained fully automatic irrigation system fed by a sunken rainwater harvest tank constructed to receive ground water from the tropical monsoon weather (Fig. 8.1).

8.2 Building Integrated Photovoltaic System

The world energy consumption is 10 terawatts (TW) per annum, and it is projected to reach 30 TW by 2050. The buildings currently consume around 40% of the world's energy, and this amount is expected to rise with further urbanization due to the growing energy demand for the new buildings. The situation should be similar in Singapore and other metropolitan cities. Some approaches have been proposed to alleviate energy demand and increase localized power generation. To eliminate the impacts on the environment in the localized power generation, a growing trend towards the use of solar energy systems, particularly photovoltaic (PV) systems has been observed. The potentials of rooftop solar photovoltaic power in urban cities have been recently investigated. Singapore has the advantage of very little seasonality of solar radiation throughout the year, which is the only abundant solar energy resource for power generation. The installation of PV panels on the rooftops of buildings is ideal as the high zenith angle (i.e., 1° North of the equator). However, the urban rooftop environment of Singapore is unlike any other, presenting challenges with a hot, humid environment that experiences significant rainfall, shadings, and overcast skies. Such conditions offer significant impacts on the performance and other deterioration of the roof PV (photovoltaic) and BIPV (Building Integrated PV) systems, resulting in degradation of the efficiency of power generation.

The Tropical Technologies Laboratory (T2 Lab) was founded in April 2016 with the sponsorship of a local real estate developer, the City Development Limited (CDL). In line with the technologies research cluster in the Department of Architecture, the T2 lab encompasses various research interests relevant to the architecture and building in the tropics. The current research objectives are directed

Fig. 8.1 Floor plan of the Tropical Technologies Laboratory (T2 Lab) indicating the eight test bed cells and facades analyzed in this study. (Adapted from: AWP Architects based on lead author's preliminary design)

toward developing passive and active building technologies appropriate to Singapore and other equatorial regions in order to achieve low/zero carbon and sustainable buildings (Fig. 8.2). Special emphasis has been given to the integration of solar panels and farming systems into modular building facades and their impact on natural cross ventilation, daylight, and thermal comfort. Building integrated photovoltaics (BIPV), both on the facades and the roof, will also be tested for the special conditions of a warm-humid equatorial climate. The lab also includes other researches focusing on building materials and technologies such as tropical timber.

Fig. 8.2 Image of the T2 Lab. On the left and right, north and west facades, respectively. (Photograph taken on June 2019)

A general model of PV integration into existing public high-rise residential buildings in Singapore was recently developed, which also presents challenges and benefits pertaining to it (Kosorić et al., 2018). In order to provide a better understanding of the whole process, the model is divided into seven basic phases detailing the role of each phase and allowing model optimization at the level of a particular phase. A systematic analysis of each phase is provided, and the problem-solving methods and/or procedures applied are discussed. The Vikor method, a multi-criteria decision-making (MCDM) method, is recommended for a comprehensive evaluation of design variants, selection of the optimal PV integration design variant, and sensitivity analysis testing the robustness of the selected design variant "optimality." The defined methodological framework is also employed to solve PV integration into an existing 12-story, slab-block public housing (HDB building). The evaluation of created design variants against aesthetic criteria was supported by a customized web-based survey and qualitative interviews that were performed in order to provide information on opinions and perceptions of local professionals regarding different roof and façade PV integration designs. The analysis of the web-based survey results offers a useful feedback for the understanding of the design approach.

Photovoltaic integrated shading devices (PVSDs) are considered a vital class of BIPV, which plays a role in generating power by transforming unwanted radiation and reducing energy consumption for cooling. A majority of the existing studies on PVSDs are in South Korea, China, Greece, and Switzerland (Zhang et al., 2018). However, there are ample solar energy resources in other nearby countries sharing similar experience of shading devices utilization to avoid excessive heating, glare while enhancing the potential of photovoltaic application. The definition and 24 types of PVSDs were clarified by Zhang et al. (2018). The multi-objective optimization study has played a crucial role in achieving the balance of building performance factors influenced by PVSD, such as cooling energy consumption, artificial lighting energy consumption, electrical energy generating, indoor visual comfort, and others, which might be a type of promising future study on PVSD.

To improve the photovoltaic (PV) power generation, the temperature control measures and optimization of BIPV systems are critical, particularly in the tropical weather. There are two categories of cooling mechanisms, i.e., passive and active cooling. Thermal behavior and airflow characteristics of the building-integrated photovoltaic (BIPV) façade have been investigated numerically. A three-dimensional model is developed based on the typical BIPV façade (Lau et al., 2020). Computational fluid dynamics (CFD) with the shear stress transport (SST) κ-omega turbulent model is used in the study. The effects of geometric configurations on the BIPV cell temperature in steady state are evaluated, including the sizes of the bottom and top openings and the depth of the return air cavity (or so-called cavity depth). When the sizes of the inlet and outlet openings are the same, the effects on the decrease of cell temperature are limited. By enlarging the bottom (inlet) opening, the impact of ventilation in the cavity behind is more significant, and the cell temperature decreases. Cavity depth is also a vital factor affecting BIPV cell temperature. The optimal cavity depth of approximately 100–125 mm is identified by Lau et al. (2020). Flow disturbance and a vortex may be observed at the bottom and top of the air cavity, respectively, as the cavity depth increases, which negatively affects the ventilation, causing these flow disturbances to increase the cell temperature.

The BIPV study aims to develop a living catalog with SERIS to demonstrate various Building Integrated Photovoltaic systems in building façade applications. Therefore, PV integration is a highly complex task requiring a dynamic, holistic approach that balances mutually interrelated, often conflicting criteria related to energy performance, economic, environmental, spatial/urban, functional, aesthetic, and social aspects. The team has found that Singapore professionals favor roof integration over façade integration, highlighting the significance of psychological and social factors in addition to the cost-efficiency ratio of PV systems. Properly educating and informing the design teams and users are essential to preclude possible opposition to façade integrations and further facilitate and accelerate PV building integration. To address these issues, allied with SERIS, T2 Lab organized various visits, seminars, and publications for promotions and communications. Some of BIPV technologies tested and demonstrated in T2 lab are listed as follows:

- REC Peak Energy Series.
- REC Blck Twinpeak.

- AGC Sudare.
- Kromatix.
- Solaxess White Module.
- CIGS.
- Digital Printing PV Panels by SERIS.

8.3   Productive Facades for Energy and Food Harvesting

The process of urbanization and densification in the tropics has led to an accelerated construction of high-rise buildings in medium-sized and large cities. Therefore, in tropical high-density urban areas, the ratio of facade versus roof surfaces is much higher than in traditional cities. This makes tropical building facades a key investigation topic in terms of thermal and visual comfort as well as the possible application of building integrated photovoltaic (BIPV) systems (Kosorić et al., 2018; Zhang et al., 2018).

On the other hand, owing to their multiple benefits, vertical greenery systems (VGS) and vertical farming (VF) technologies have been implemented in a growing number of existing and new buildings all around the world (Beacham et al., 2019; Despommier, 2010). VF is applied in rooftops and indoors in most cases which, despite its logistic and pest-control benefits, requires large amount of energy and the use of highly valuable indoor space. Conversely, the use of facades for farming takes advantage of the free energy provided by the sun and does not affect higher-profit indoor activities. Several studies have shown that building facades in the Equator receive enough sunlight to generate electricity by installing PV modules as shading devices (Tablada & Zhao, 2016) and for growing leafy vegetables (Song et al., 2016; Tablada & Zhao, 2016).

8.3.1   Design Optimization and Application of Productive Facades at T2 Lab

One of the investigations carried out at the Tropical Technologies Laboratory (T2 Lab) focused on the concept and assessment of the Productive Facade (PF) integrating photovoltaic (PV) and farming systems. Eight PFs were designed and installed at four cardinal orientations corresponding to eight testing cells. Two types of facades representing typical Housing Development Board (HDB) building facades were assessed for each orientation: balcony facade and window facade (Fig. 8.3). The design of the eight facades was conducted by using 3D simulation algorithms and by applying a multi-criteria decision-making (MCDM) process. VIKOR optimization method (Opricovic & Tzeng, 2004, 2007) was used to find the best compromise among five performance indicators: (1) food production potential, (2)

Fig. 8.3 Productive facades – north and east – at the NUS-CDL T2 Lab in January 2019 at 9 AM. (Image Credit: Abel Tablada)

potential electricity generation, (3) indoor daylight, (4) solar heat gain, and (5) view angles from the interior. Other design aspects related to cost, accessibility, and aesthetics were also considered. Each performance indicator was assessed according to the amount, position, size, and tilt angle of the PV modules acting as shading devises on top of each facade. Also, the position and separation between planters were considered.

Once the optimal design variants were selected, some minor adjustments were done to the final installation of the eight PF at the lab based on the site conditions and the results of an online survey involving 100 PV experts and architects (Tablada et al., 2020) as well as a door to door social acceptance survey among HDB residents (Kosorić et al., 2019). The final optimal prototypes were adjusted after considering the survey results. The position and dimensions of all PV modules assure acceptable indoor visual and thermal conditions while allowing the required sunlight on the crops. The copper indium gallium selenide (GIGS) PV modules located on north and south facades were selected for having a good response under non-clear skies – a typical condition of Singapore – and taking into account that half of the year each facade would not receive direct solar radiation. On the other hand, monocrystalline silicon PV modules were installed on the east and west facades as they are more efficient under direct solar radiation. Regarding the VF, a tropical-adapted variety of lettuce was selected among several commonly cultivated leafy vegetables in Singapore. Apart from being considered as one of the preferred vegetables in Singapore, lettuces have also short harvest cycles and a moderate light requirement. However, other leafy vegetables such as Chinese cabbage, water spinach (Kang Kong) and spices can also be cultivated in PF systems in Tropical regions. Lettuces were cultivated in six rounds from December 2018 to June 2019. Fertilizer was applied twice per month on a soil mix of coco peat and perlite. An automated drip irrigation system with micro-spread heads on each plant was used and activated three times per day for 2 minutes.

8.3.2 Measurement Results

Measurement data were collected on the PFs and inside the 8 cells from 78 sensors which monitored air and surface temperatures, relative humidity, solar irradiance, light intensity, water consumption, and electric power production. Wind conditions and the effect of the VF system on natural ventilation inside the cells were also analyzed by Yuan et al. (2019). After adjustment and calibration, data collection related to PV modules started from April 2019. Vegetable production was quantified by weighting fresh lettuces. There were six rounds of crops from December 2018 to June 2019.

Regarding the electricity generation from PV modules, the results from April, August, October (2019), and January (2020) are shown in Fig. 8.4. Electricity generated by PV modules on the east and west facades were from 1.7 to 2.2 times higher than on the south and north facades. Considering a typical HDB household with both north and south facades (20 m in total), around 30% of energy demand can be supplied by using PV modules as shading devices. Those buildings in which main facades are oriented closer to east and west are expected to satisfy a higher percentage of energy demand.

Regarding the production of vegetables, lettuces grew well on all T2 lab facades as shown in Fig. 8.5. However, total yield was declining except for the last month when new soil was added. This indicates the impact soil quality has on vegetable yield. East and north facades have the greatest performance, producing 902 g and 828 g of lettuce. South and west facades have a total yield of 763 g and 550 g, respectively. The amount produced represents 55–103% of the average leafy vegetable consumption of a 4-member household in Singapore (ca. 16 kg per year).

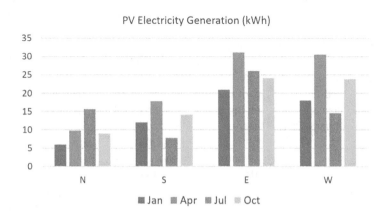

Fig. 8.4 Electricity generation from PV modules per facade orientation for the months of January (2020), April, July, and October (2019)

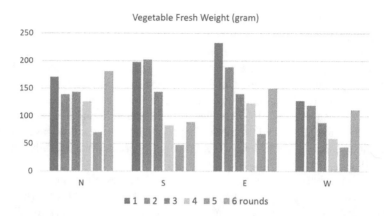

Fig. 8.5 Weight of fresh lettuces cultivated at the T2 Lab on each of the six rounds for the four facade orientations

8.4 Natural Ventilation and Vertical Farming

(Excerpts from an original paper published in Energy and Buildings, 185, pp. 316–325) Yuan C, Shan RQ, Adelia AS, Tablada A, Lau SK, Lau SSY, 2019, Effects of Vertical Farming on Building Cross Natural Ventilation at Urban Residential Areas (copyright: Yuan Chao)

8.4.1 Introduction

Urban farming, often defined as growing food within and around cities, has become a global trend over the past decades. It arises as a response to the food safety and land resource issues, sustainable living, and environmental degradation due to rapid urbanization (Shamshiri et al., 2018; Smit et al., 1996; Valley & Wittman, 2018), and many of the benefits of farming on urban living have been recognized by international organizations, such as World Food and Agriculture Organization (FAO) and UN-habitat (FAO, 2014; UN-Habitat, 2014). Vertical farming, i.e., growing plants in stacked layers, is one of the most common types of urban farming, especially at high-density cities with limited land resources such as Singapore and Hong Kong (Al-Kodmany, 2018; Pinstrup-Andersen, 2018). Similar with vertical greenery (Wong et al., 2009; Wong et al., 2010), vertical farming can benefit urban heat island mitigation, thermal insulation and shade, and noise pollution reduction (Coma et al., 2017; Köhler, 2008; Manso & Castro-Gomes, 2015; Pérez et al., 2011; Tablada et al., 2018). Contrary to vertical greenery that is normally planted on a solid wall, vertical farming is situated at the porous part of buildings (such as balcony as shown in Fig. 8.1), which is very important to the building natural ventilation performance. It has been shown that the vegetation could worsen the natural ventilation due to additional drag force on air flow (Gromke, 2011; Yuan et al., 2017). Therefore, passive design strategies to mitigate the negative impact of vertical farming on natural ventilation, especially at high-density tropical/subtropical cities, are desperately needed, since the thermal discomfort and stagnant air flow are environmental issues at such cities (Cheng et al., 2011; Ng et al., 2011; Yuan et al., 2016; Yuan & Ng, 2012). This study provides important understandings on vertical farming to enable architects to make the evidence-based decision in the architectural design.

8.4.2   Methodology

In this study, the effect of vertical farming on the natural ventilation was investigated using Computational Fluid Dynamics (CFD) simulations with ANSYS Fluent. Prior to modelling the façade, a validation test was implemented against a wind tunnel experiment in a street canyon with avenue-like tree planting. A validated RSM turbulence model was used to successfully simulate the dynamics of air flow and emission dispersion with tree planting as a porous medium in the street canyon. In the parametric study of the façade with vertical farming, the settings of the porosity and pressure loss coefficient were the same with the ones applied in the validation test. As shown in Fig. 8.6, there are six designs with respect to the block ratio of the façade and the arrangement of vegetable and vegetable species in the parametric study.

8.4.3   Results and Discussion

Taking Case 3a as an example, the velocity vectors near the facade are given in Fig. 8.7 to show the effect of vegetable to the flow field. It is shown in Fig. 8.6 that the wind speed accelerates in the gap between vegetable rows and decreases when passing the vegetable due to the drag force of vegetable on air flow. To quantitatively compare the natural ventilation efficiency of the façade with vegetable, the horizontal profiles along incoming wind direction for selected test lines located in the vertical plane were plotted to collect the wind speed data.

The quantitative analysis indicates that the block ratio of vegetable plays an important role in the ventilation performance. When three-fourths of the façade is covered by vegetable, i.e., block ratio of 0.75, the ventilation of the façade is very limited. For the façade with a block ratio of 0.5, half of incoming air flow can reach the near field downstream of the façade to promote indoor thermal comfort. Therefore, in the design process of the vertical farming, the block ratio of vegetable should be carefully determined regarding the ventilation performance. Furthermore, the natural ventilation with the same block ratio of vegetable could also be improved by appropriately modifying vegetable arrangement and vegetable species. Regarding

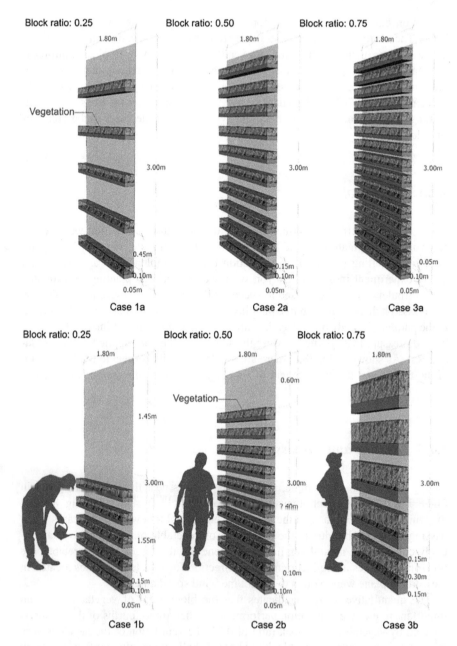

Fig. 8.6 Schematic of parametric models

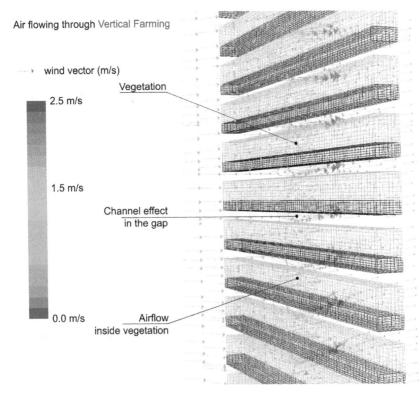

Fig. 8.7 Velocity vector in the vertical plane in the middle of the domain, taking Case 3a as an example

to the vegetable arrangement, it is important to notice that the upper level wind speed (both minimum and average wind speed at H = 2.5 m) at Case 2b has not been significantly increased compared with Case 2a, and the lower level wind speed (both minimum and average wind speed at H = 2 m, 1.5 m, 1 m, and 0.5 m) is much lower than that of Case 2a. It is because the gaps at lower level was reduced from 0.15 m to 0.10 m, which significantly decreases the leeward wind speed, and at the same time, the gap at high level is only 0.6 m that is not wide enough to avoid the impact of vegetable on air flow. On the other hand, the vegetable arrangement can be modified by choosing different vegetable species, as shown in Fig. 8.1, Case 3b, in which the higher vegetable is chosen. With the same vegetable block ratio, the number of rows was reduced from 15 to 5 rows, and the gaps between rows become wider, 0.15 m. The new arrangement performs slightly better than the baseline case, i.e., higher minimum wind speed and similar averaged wind speed. The minimum wind speed was increased to 0.7 m/s, which could be helpful in improving thermal comfort.

While vertical farming can promote sustainable living and mitigate urban heat island, vegetable could also worsen the cross natural ventilation due to additional drag force on air flow. This study investigated the impact of various vertical farming configurations on natural ventilation performance using CFD simulation and provided the important understandings to enable architects to make the evidence-based decision in the architectural design.

Acknowledgments Our gratitude to the following for contributions in one way or another – fund donation, donor of equipment and materials, research collaborators, supporter for technical advice, land and project establishment, and the research team (arranged in alphabetical order),

AVA -Horticulture Department as collaborator.

AWP Architects Pte. Ltd. as Architect, RP.

Building Control Authority BCA for funding the tropical timber research which engages with the Lab as test-bed for onsite testing.

City Developments Limited – for their generous unconditional donation.

LIAN HO LEE Construction Pty. Ltd. as Specialist Contractor.

NUS Faculty of Science, Department of Biological Sciences as collaborator.

NUS Office of Estate Development OED as Project Manager.

NUS Office of Housing Services OHS who provides the land for the laboratory.

NUS School of Design & Environment for establishment and funding.

NUS- Solar Energy Research Institute of Singapore (SERIS), for donation and fabrication of the GIGS and c-Si modules respectively. SERIS is also a Sponsor and Collaborator.

WBG (SG) Pte. Ltd. as Main Contractor.

The following as Sponsor: AGC, Balkan Energy AG, KROMATIX, REC, ROTHOBLAAS, SERIS, SOLAXESS, TERMIMESH, UNISEAL.

Following individuals: Ayu Sukma Adelia, Ian K. Chaplin, Huang Huajing, Vesna Kosoric, Shi Xuepeng, Thomas Reindhl, Veronika Shabunko, Shan Ruiqin, Hugh Tan, Song Shuang, and Zhang Qianning.

References

Al-Kodmany, K. (2018). The vertical farm: A review of developments and implications for the vertical city. *Buildings, 8*(2), 24.

Beacham, A. M., Vickers, L. H., & Monaghan, J. M. (2019). Vertical farming: A summary of approaches to growing skywards. *The Journal of Horticultural Science and Biotechnology, 94*(3), 277–283.

Benton, C. C., & Kwok, A. G. (1995). The Vital Signs Project: Work in Progress.

Building Research Establishment BRE, UK(2020), retrieved from https://www.bregroupcom/about-us/our-history

Cheng, V., Ng, E., Chan, C., & Givoni, B. (2011). Outdoor thermal comfort study in a subtropical climate: A longitudinal study based in Hong Kong. *International Journal of Biometeorology.* https://doi.org/10.1007/200484-010-0396-z

Coma, J., Pérez, G., de Gracia, A., Burés, S., Urrestarazu, M., & Cabeza, L. F. (2017). Vertical greenery systems for energy savings in buildings: A comparative study between green walls and green facades. *Building and Environment, 111*, 228–237.

Despommier, D. (2010). *The vertical farm: Feeding the world in the 21st century.* Macmillan.

FAO. (2014). Urban agriculture. Retrieved from http://www.fao.org/3/a-i3696e.pdf

Gromke, C. (2011). A vegetation modeling concept for building and environmental aerodynamics wind tunnel tests and its application in pollutant dispersion studies. *Environmental Pollution, 159*(8–9), 2094–2099.

Institute., I. W. B. (2020). Retrieved from https://www.wellcertified.com/certification/v1/standard

Köhler, M. (2008). Green facades—A view back and some visions. *Urban Ecosystem, 11*(4), 423.

Kosorić, V., Huang, H., Tablada, A., Lau, S.-K., & Tan, H. T. (2019). Survey on the social acceptance of the productive façade concept integrating photovoltaic and farming systems in high-rise public housing blocks in Singapore. *Renewable and Sustainable Energy Reviews, 111*, 197–214.

Kosorić, V., Lau, S.-K., Tablada, A., & Lau, S. S.-Y. (2018). General model of Photovoltaic (PV) integration into existing public high-rise residential buildings in Singapore – Challenges and benefits. *Renewable and Sustainable Energy Reviews, 91*, 70–89. https://doi.org/10.1016/j.rser.2018.03.087

Kwok, A. G., Benton, C. C., & Burke, B. (1997). The vital signs project: Dissemination activities American Solar Energy Society, Boulder, Co, USA..

Lau, S.-K., Zhao, Y., Lau, S. S. Y., Yuan, C., & Shabunko, V. (2020). An investigation on ventilation of building-integrated photovoltaics system using numerical modeling. *Journal of Solar Energy Engineering, 142*(1). https://doi.org/10.1115/1.4044623

Manso, M., & Castro-Gomes, J. (2015). Green wall systems: A review of their characteristics. *Renewable and Sustainable Energy Reviews, 41*, 863–871.

McArthur, J., & Powell, C. (2020). Health and wellness in commercial buildings: Systematic review of sustainable building rating systems and alignment with contemporary research. *Building and Environment, 171*, 106635.

Ng, E., Yuan, C., Chen, L., Ren, C., & Fung, J. C. H. (2011). Improving the wind environment in high-density cities by understanding urban morphology and surface roughness: A study in Hong Kong. *Landscape and Urban Planning, 101*(1), 59–74.

Opricovic, S., & Tzeng, G.-H. (2004). Compromise solution by MCDM methods: A comparative analysis of VIKOR and TOPSIS. *European Journal of Operational Research, 156*(2), 445–455.

Opricovic, S., & Tzeng, G.-H. (2007). Extended VIKOR method in comparison with outranking methods. *European Journal of Operational Research, 178*(2), 514–529.

Pérez, G., Rincón, L., Vila, A., González, J. M., & Cabeza, L. F. (2011). Green vertical systems for buildings as passive systems for energy savings. *Applied Energy, 88*(12), 4854–4859. https://doi.org/10.1016/j.apenergy.2011.06.032

Pinstrup-Andersen, P. (2018). Is it time to take vertical indoor farming seriously? *Global Food Security, 17*, 233–235.

Preiser, W. F., Davis, A. T., Salama, A. M., & Hardy, A. (2014). *Architecture beyond criticism: Expert judgment and performance evaluation*. Routledge.

Preiser, W. F., & Schramm, U. (1997). Building performance evaluation. In *Time-saver standards for architectural design data* (Vol. 7). McGraw-Hill.

Shamshiri, R. R., Kalantari, F., Ting, K. C., Thorp, K. R., Hameed, I. A., Weltzien, C., et al. (2018). Advances in greenhouse automation and controlled environment agriculture: A transition to plant factories and urban agriculture. *International Journal of Agricultural & Biological Engineering, 11*(1), 1–22. https://doi.org/10.25165/j.ijabe.20181101.3210

Smit, J., Nasr, J., & Ratta, A. (1996). *Urban agriculture: Food, jobs and sustainable cities* (Vol. 2, pp. 35–37). United Nations Development Programme (UNDP).

Song, A., Lu, L., Liu, Z., & Wong, M. S. (2016). A study of incentive policies for building-integrated photovoltaic technology in Hong Kong. *Sustainability, 8*(8), 769.

Tablada, A., Kosorić, V., Huang, H., Chaplin, I. K., Lau, S.-K., Yuan, C., & Lau, S. S.-Y. (2018). Design optimization of productive Façades: Integrating photovoltaic and farming systems at the Tropical Technologies Laboratory. *Sustainability, 10*(10), 3762.

Tablada, A., Kosorić, V., Huang, H., Lau, S. S., & Shabunko, V. (2020). Architectural quality of the productive façades integrating photovoltaic and vertical farming systems: Survey among experts in Singapore. *Frontiers of Architectural Research 9*(2), 301–308..

Tablada, A., & Zhao, X. (2016). Sunlight availability and potential food and energy self-sufficiency in tropical generic residential districts. *Solar Energy, 139*, 757–769.

Turpin-Brooks, S., & Viccars, G. (2006). The development of robust methods of post occupancy evaluation. *Facilities*. https://doi.org/10.1108/02632770610665775

UN-Habitat. (2014). Integrating urban and peri-urban agriculture into city-level climate change strategies. Retrieved from https://unhabitat.org/integrating-urban-and-peri-urban-agriculture-into-city-level-climate-change-strategies-june-2014

Valley, W., & Wittman, H. (2018). Beyond feeding the city: The multifunctionality of urban farming in Vancouver, BC. *City, Culture and Society*. https://doi.org/10.1016/j.ccs.2018.03.004

White, A. H. (1826) The Bartlett, architectural pedagogy and Wates house – An historical study. *Opticon, 16*, 26, 1–19, https://doi.org/10.5334/opt.ci

Wong, N. H., Tan, A. Y. K., Chen, Y., Sekar, K., Tan, P. Y., Chan, D., et al. (2010). Thermal evaluation of vertical greenery systems for building walls. *Building and Environment, 45*(3), 663–672.

Wong, N. H., Tan, A. Y. K., Tan, P. Y., & Wong, N. C. (2009). Energy simulation of vertical greenery systems. *Energy and Buildings, 41*(12), 1401–1408.

Yuan, C., & Ng, E. (2012). Building porosity for better urban ventilation in high-density cities – A computational parametric study. *Building and Environment, 50*, 176–189.

Yuan, C., Ng, E., & Norford, L. (2017). A semi-empirical model for the effect of trees on the urban wind environment. *Landscape and Urban Planning, 168*, 84–93.

Yuan, C., Norford, L., Britter, R., & Ng, E. (2016). A modelling-mapping approach for fine-scale assessment of pedestrian-level wind in high-density cities. *Building and Environment, 97*, 152–165.

Yuan, C., Shan, R., Adelia, A. S., Tablada, A., Lau, S. K., & Lau, S. S.-Y. (2019). Effects of vertical farming on natural ventilation of residential buildings. *Energy and Buildings, 185*, 316–325.

Zhang, X., Lau, S.-K., Lau, S. S. Y., & Zhao, Y. (2018). Photovoltaic integrated shading devices (PVSDs): A review. *Solar Energy, 170*, 947–968.

Chapter 9
Field Study on Thermal Comfort and Adaptive Behaviors of University Students in the Cold Climate Zone

Shimeng Hao, Zhonghua Gou, Yufei Zou, and Xiaoshan Xing

9.1 Introduction

Nowadays, people choose to stay indoors most of the time; one study suggested that the time is as high as 90%. For students, they spend more time in the classrooms because of their study. As a result, individuals start to pay attention to the comfortable indoor thermal conditions in these buildings (Zomorodian et al., 2016). Some research reported that a poor classroom indoor environment had an impact on students' daily behavior and individuals' health, such as absenteeism, decreased academic performance, and poor health symptoms (Eide et al., 2010; Mendell et al., 2013). And other studies have shown that it is the thermal environment and air quality inside classrooms that correlate strongly with student learning and productivity, which is exactly why educational buildings need to provide a good environment for students to improve their performance and well-being (Haverinen-Shaughnessy et al., 2015). Therefore, appropriate indoor conditions are necessary for students.

People are also proposed to actively adapt to their environment by three influence factors: physiological, psychological, and behavioral adaptabilities. An adaptive model raised by Dear and Brager presumed that human does not passively accept their environment. They adapt to their environment as active participants through multiple feedback loops and interactions between occupant and environment (Brager & de Dear, 2000; de Dear & Brager, 1998, 2002). A study taken in Chongqing found that students' acceptable temperature was dependent on indoor temperature. The range was wide when the temperature was neutral in the class-

S. Hao · Y. Zou · X. Xing
School of Architecture and Urban Planning, Beijing University of Civil Engineering and Architecture, Beijing, China

Z. Gou (✉)
School of Urban Design, Wuhan University, Wuhan, China

© The Author(s), under exclusive license to Springer Nature Switzerland AG 2021   111
S. S. Y. Lau et al. (eds.), *Design and Technological Applications in Sustainable Architecture*, Strategies for Sustainability, https://doi.org/10.1007/978-3-030-80034-5_9

room while narrow when it was overheated or overcooled (Yao et al., 2010). The students in Shaanxi Province showed better tolerance to cooler temperatures in classrooms, compared with the warmer indoor temperatures in winter. Results from this survey also mentioned that indoor design temperature should be taken into full consideration, because the high temperature indoors not only leads to the waste of energy but also affects the comfort of students (Wang, Yan, et al., 2017a).

The increasing comfort levels and time spent in buildings cause building energy consumption at a high level (Pérez-Lombard et al., 2008). Beijing is in the Cold Climate Zone of China, where it should meet the requirements of winter heat preservation. The space heating lasts about 5 months a year, and energy consumption is huge. Compared with the temperature setting, a large number of heating hours caused a lot of energy consumption (Guerra-Santin & Itard, 2010). In the energy consumption of buildings, some researchers pointed out that education buildings occupy a large proportion (Rupp et al., 2018).

In addition to heating and air conditioning changes in indoor temperature, various "adaptive opportunities" were used to adjust individuals' indoor thermal environment (Rijal et al., 2018). Students at school take many measures to adapt to changes in indoor temperatures, such as adding clothes in winter, opening windows, turning on ventilation in summer, etc. These all show their ability to adapt to changes in indoor temperature (Jindal, 2018). A study taken in university classrooms in Zunyi reported when the room temperature rises, the average of students' clothing insulation generally decreases (Liu et al., 2019). Some researchers that conducted a field investigation in non-air-conditioned classrooms in Eindhoven found that, over a year, the thermal resistance of the students' clothing varied from 0.9 in the cold season to 0.3 in the hot season, with the greatest change occurring in the middle of the season (terMors et al., 2011). In Nepal's temperate climatic region, only when the outdoor temperature was above 30 did students' clothing adaptive behavior become more pronounced (Shrestha et al., 2021). For adaptive behavior, much of the focus is on individual self-regulation, such as adding or removing clothing. There is not much research on the thermal adaptation of group living.

There are many differences in the performance of individual thermal comfort. Therefore, when carrying on the architectural design, this point needs to be fully considered by the designer (Liu et al., 2019). The two most important reasons for individual differences are gender and age (Wang et al., 2018). Some scholars concluded that in the same high-temperature environment, women and the elderly are less tolerant of high temperature than men and young people. In cool conditions, they usually feel cooler than the current temperature (Karjalainen, 2012). In a study in Australia, researchers found under the same thermal environmental conditions, compared with adults, students are more inclined to lower thermal temperature (de Dear et al., 2015). The impact of gender gets more attention when we focus on college campuses where people of a similar age are. A field investigation in universities in the U.K. showed that women felt comfortable at a higher temperature than men. The investigation had found that women wore more clothing to keep warm when their comfort temperatures were similar (Jowkar et al., 2020). And a field investigation in Nepal shown that female students' comfort temperatures are higher than male students (Shrestha et al., 2021).

Based on the above previous scholars' research on this aspect, we conducted a study at Beijing University of Civil Engineering and Architecture to investigate students' thermal comfort and adaptation when they stayed in classrooms. For architectural design students, they need a more positive indoor environment to maintain their creativity. This study is taken in the classroom as a unit to explore the form of thermal comfort regulation in group life and the difference in group and individual regulation. The field study started from a non-heating period to a heating period. Therefore, the thermal adaptability during the heating transition period as well as the energy conservation based on human thermal comfort was also discussed in this article.

9.2   Material and Methods

9.2.1   Geographical and Climatic Conditions

The field research was done in Beijing, which is located in Northern China (39° 54′ 24″ N, 116° 23′ 51″ E). Beijing's climate is categorized as "Monsoon-influenced Humid Continental Climate," with very hot, humid summers and brief yet cold, dry winters, according to the climatological method of classification. In spring and autumn, i.e., transitional seasons, it always has minimal precipitation. Dust storms blow from the Gobi Desert over the Mongolian steppe in the spring, followed by rapid warming, while autumn is cool and less windy. The monthly average daily temperature in January is 2.9 °C, while it is 26.9 °C in July. Extremes have ranged from 27.4 °C on February 22, 1966, to 41.9 °C on July 24, 1999 (unofficial average of 42.6 °C on June 15, 1942) since 1951 (https://en.wikipedia.org/wiki/Beijing#Climate, n.d.). According to the classification of building climate zone in China (GB, 2016), Beijing is categorized as the Cold Climate (Fig. 9.1).

9.2.2   Research Time

The field survey was conducted from October 15, 2018, to January 12, 2019, including a transitional season of autumn to winter, which was also defined as a non-heating period and a heating period. The daily mean outside air temperature fluctuates but generally shows a downward trend during this period. There was a rapid fall from December 3, 2018, and it dropped to −10 °C for the first time on December 7, 2018. As is shown in Fig. 9.2, the research period was divided into three phases, and they

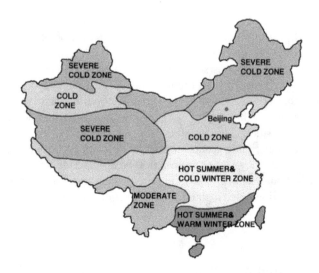

Fig. 9.1 The building climate demarcation of China and the location of Beijing

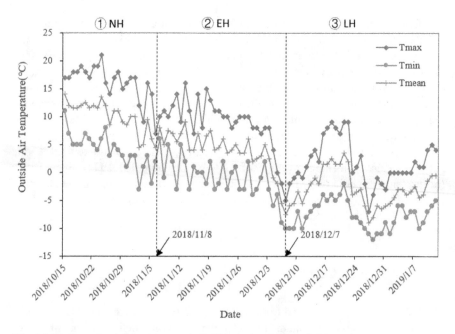

Fig. 9.2 Daily outdoor air temperature in Beijing during the investigation

were respectively termed as non-heating period (NH), early heating period (EH), and late heating period (LH). The reasons are considered as follows:

1. To make each period contain the same number of days.
2. The outdoor temperature drops to $-10\ ^\circ$C for the first time, and the environmental factors affecting human thermal comfort have great changes.
3. People have a certain adaptation period to the environmental changes after heating.

The numbers of investigated days were, respectively, 20 days in phase 1, 20 days in phase 2, and 19 days in phase 3(not counting the weekends and holidays).

9.2.3 Investigated Building and Classrooms

From October 2018 to January 2019, a field study was carried out in 19 classrooms of an educational building, which was mixed-mode in fall and naturally ventilated in winter. There are ten classrooms on the first floor, among which the area of the seminar room was smaller, and other classrooms are the same. There are nine classrooms, one large seminar room, and two small seminar rooms on the second floor. Figures 9.3 and 9.4 show respectively the layout of different floors and pictures of the target building, which was built in the 1940s and renovated in 2017. Casement is the main windows opening type. The occupants were seating for the most time, doing their normal learning tasks.

9.2.4 Sample Selection

One hundred fifty-eight university students who volunteered as the participants were tracking the survey. Nineteen classrooms were surveyed on campus in Beijing. The female-to-male ratio is 3:2. All of the participants had fully adapted to the local climate and campus life for they had been living in Beijing for more than 2 years. The participants' basic demographic and physical information including age, weight, height, and BMI is shown in Table 9.1. As can be seen from the table, the average BMI of participants was in the normal region (18.5–24.9), which can guarantee the accuracy of their investigation data.

9.2.5 Equipment and Questionnaire

There were two primary parts of the field study. The first part was objective data measurement, including onsite observation of the indoor thermal parameters. In the second part, the occupants filled out the thermal comfort questionnaires with their subjective responses to the indoor environment.

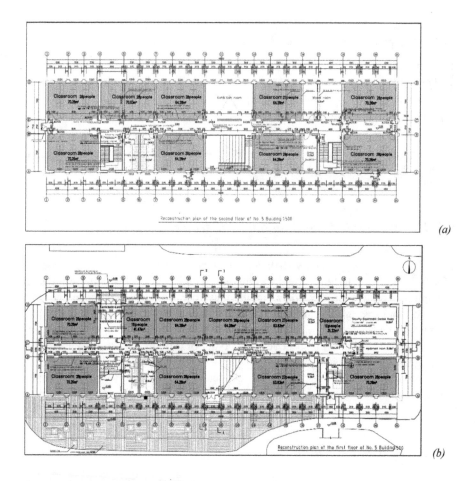

Fig. 9.3 Layout of the studied educational building in Beijing of (**a**) the first floor and (**b**) the second floor

During the study, objective measurements and subjective questionnaires were obtained at the same time. The technique for conducting onsite investigations was identical to that used in previous field studies (López-Pérez et al., 2019; Wang, Ning, et al., 2017b). The environmental measurements, including indoor air temperature, globe temperature, airspeed, and relative humidity, were measured onsite respectively by the instruments shown in Table 9.2. The measurement points were chosen according to the ASHRAE 55–2017 (ASHRAE, 2017) and ISO standard 7730 (ISO, 2005). The research instruments were all set at a distance of more than 2 meters from the participants. Also, they were kept away from direct sunlight, cold windows, walls, and other heat appliances like computers or desk fans. Every day during the study, the questionnaire had to be completed independently on the internet. Clothing, activity level, thermal sensation and comfort, thermal expectation, and other factors were all discussed. Table 9.3 summarizes the voting scale.

(a) the first floor and (b) the second floor

Fig. 9.4 The pictures of the studied educational building in Beijing

Table 9.1 Participants' demographics and physical characteristics

		Age distribution	Weight (kg)	Height (m)	[a]BMI (kg/m²)
158(68 male/ 90 female)	Average	18.46	61.04	1.69	21.1
	Standard deviation	0.674	14.556	0.086	3.8
	Minimum	16	36.5	1.47	14.3
	Maximum	21	125.0	1.94	38.6

[a]Body Mass Index (BMI) = Weight (kg)/ (Height (m))²

Table 9.2 Detailed information on test instruments

Indices	Test instrument	Instrument	Measurement range	Measurement accuracy
Indoor air temperature	Thermo-recorder	MX1101	−20 ~ +70 °C	±0.21 °C (0 ~ 50 °C)
Relative humidity	Thermo-recorder	MX1101	1% ~ 90% RH	±2% (20% ~ 80%)
Globe temperature	Globe thermometer	TESTO480	−100 ~ +400 °C	0.01 °C
Air velocity	Anemograph	TESTO480	0.1 ~ +15 m/s	0.01 m/s

Table 9.3 The coding scheme used in the subjective questionnaire

Measuring scale	Thermal sensation vote	Thermal preference	Thermal acceptance	Humidity feeling	Humidity preference	Air movement feeling	Air movement preference
3	Hot		Unacceptable	Very humid		Very high	
2	Warm		Just unacceptable	Humid		High	
1	Slightly warm	Cooler	Just acceptable	Slightly humid	More humid	Slightly high	Stronger
0	Neutral	No change	Acceptable	Neither humid nor dry	No change	Neither high nor low	No change
−1	Slightly cool	Warmer		Slightly dry	Drier	Slightly low	Slighter
−2	Cool			Dry		Low	
−3	Cold			Very dry		Very low	

9.2.6 The Predicted Mean Vote (PMV)

The effective temperature (ET), the normal effective temperature (SET), and the operative temperature (T_{op}) are some of the indices that can be used to measure the thermal state (Fanger, 1972). The thermal comfort parameters considered both the objective measurements and the subjective preference of the reviewed. The design values for the standard of comfort were provided by the standards ISO 7730 (ISO, 2005) and EN 15251 (CEN, 2007). These figures were derived from the operating temperatures at the schools, which were calculated using heat balance and thermal comfort models.

As a result, T_{op} was used as a thermal comfort index to quantify the thermal sensation. Thus, T_{op} was used to quantify the thermal sensation, i.e., as a thermal comfort index. The operative temperature can be calculated as shown in Eq. 9.1. The mean radiant temperature (T_r) was calculated based on the air temperature, globe temperature, and air velocity, which is seen in Eq. 9.2.

$$T_{op} = T_r + \frac{T_a \times \sqrt{10V_a}}{1 + \sqrt{10V_a}} \tag{9.1}$$

$$T_r = T_g + 2.73\sqrt{V_a}\left(T_g - T_a\right) \tag{9.2}$$

where T_r is mean radiation temperature, the T_g is the black globe temperature, he T_a is indoor air temperature, and V_a is air velocity.

Based on experimental data, the PMV was calculated by CBE Thermal Comfort Tool online. The seven-point comfort scale can be used by the PMV model to predict a mean judgment of the thermal environment from a large group of people.

9.3   Results and Analysis

9.3.1   Environmental Characteristics of the Surveyed Classrooms

Table 9.4 shows the indoor environmental parameters for each of the three phases. The average indoor air temperature during the study period was 22.9 °C, with a range of 17.8 °C to 28.7 °C. During the NH phase, Ta recorded very low temperatures, as high as 17.8 ° C, which may be due to the few temperature control regulations. The Ta in the EH phase was higher than those in the other two phases; the mean differences were 2.07 °C for the NH phase and 1.16 °C for the LH phase. The T_g of the EH phase was also higher than that of the other two phases. The air velocity, clothing thermal insulation, and activity level resulted without considerable differences among the three phases. While the RH of the three phases shows an obvious decrease as time goes by. The indoor air temperatures were generally close to the upper limit of thermal comfort 24 °C in winter according to ASHRAE 55 Standard (ASHRAE, 2017). The relative humidity was within 60%, and airspeed was within 0.60 m/s in each phase, which met the requirements in ASHRAE 55 Standard (ASHRAE, 2017).

9.3.2   Result of Subjective Thermal Comfort

9.3.2.1   Frequencies of Thermal Sensation, Thermal Preferences, and Acceptability

Figure 9.5 illustrates the frequency distribution of votes through the TSV, TP, and TA scales across all three stages. In these three phases, most students felt a "neutral" sensation. When heating was offered in winter, the number of felt "slightly cool" and "cold" students dropped significantly by 10.06%. However, the number of students who felt "slightly warm" increased 11.52% during the EH phase compared to the NH phase, and the figure was reduced by 4.99%. The number of students who voted for the "warm" sensation showed the same trend in the three phases. This indicates that in the EH, students felt uncomfortable with the high-temperature indoor environment. This is because students are already thermally adapted to the lower room temperature when in the NH phase.

At the same time, people's thermal preference also tends to be "slightly warmer," after providing heating, which may be due to the decline of outdoor temperature and people's adaptation to the warm indoor temperature. At each phase, a host of students' choices for the current thermal environment were "acceptable." In the NH phase, the comfort temperature was set at 22.1 ± 1.2 °C. In the EH phase, the comfort temperature was set at 24.0 ± 1.6 °C, and in the LH phase, the temperature was

Table 9.4 Indoor environmental parameters in different phases

Parameters	NH				EH				LH			
	Min.	Max.	Mean	Standard deviations	Min.	Max.	Mean	Standard deviations	Min.	Max.	Mean	Standard deviations
N	763				287				311			
T_a(°C)	17.80	25.90	22.23	1.23	20.50	28.10	24.30	1.57	18.40	28.70	23.14	1.86
T_g(°C)	18.00	27.10	22.16	1.29	20.50	28.30	24.32	1.70	19.30	28.00	23.40	1.53
V_a (m/s)	0.00	0.60	0.05	0.05	0.00	0.30	0.09	0.06	0.00	0.55	0.05	0.05
RH (%)	20.10	59.30	42.26	9.65	9.50	39.70	22.56	7.59	6.10	19.20	11.51	2.83
Clo (clo)	0.495	1.86	0.92	0.25	0.50	2.08	1.08	0.31	0.5	2	1.06	0.31
Met (met)	0.7	5	1.453	0.774	0.7	5	1.889	0.932	0.7	5	1.770	1.123

N number of samples, T_a Indoor air temperature (°C), T_g black globe temperature (°C), V_a indoor air velocity (m/s), RH relative humidity (%), Clo clothing insulation (clo), Met metabolic equivalent (met)

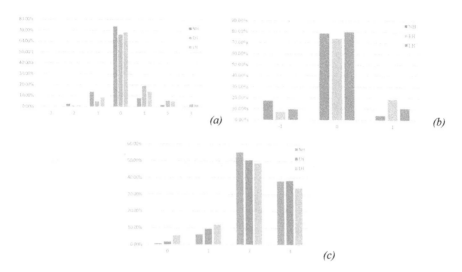

Fig. 9.5 Frequencies of (**a**) the TSV (thermal sensation votes), (**b**) the TP (thermal preference), and (**c**) the TA (thermal acceptability)

at 23.0 ± 1.8 °C. This shows that students' acceptance of temperature will change with different temperature environments and psychological feelings.

9.3.2.2 Relative Humidity Feeling

Some students felt dry in winter. Especially after heating, more and more students voted for a "slightly dry" sensation. In the LH phase, the number has occupied 36.1% of the students tested, and the other 10.0% of them voted for a "dry" sensation. Meanwhile, after providing heating, there are no students who voted for a "humid" sensation. Heating leads to lower indoor humidity in classrooms, which makes students more likely to prefer more humid environments. With the students' adaptation to the environment, in the later stage of heating, the students' adaptability to indoor humidity has slightly improved. Therefore, in addition to ensuring heating, we should take other measures to provide a suitable humidity environment for classrooms (Fig. 9.6).

9.3.2.3 Air Velocity

After heating, the students felt that the indoor air velocity was lower. Compared with the NH phase, there was a 13.42% increase in the number of students who experienced low air velocity in the EH phase. This value was not much different between the LH phase and the EH phase. However, 34.84% of the students expected a higher air velocity in the EH phase, and this value dropped to 24.52% in the LH

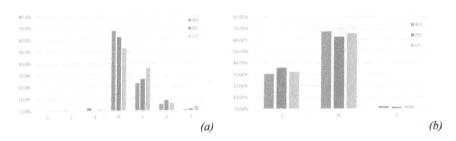

Fig. 9.6 Frequencies of (**a**) the HF (humidity feeling) and (**b**) the HP (humidity preference)

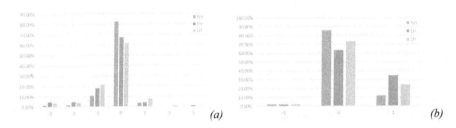

Fig. 9.7 Frequencies of (**a**) the AMF (air movement feeling) and (**b**) the AMP (air movement preference)

phase. In addition to the increased indoor temperature slowing down the flow of air, the cold temperatures outside also make opening the window adjustment less helpful (Fig. 9.7).

9.3.3 Thermal Adaptive Behaviors

In this field investigation, the participants were asked a question about the thermal adaptive behaviors: "Since this morning, for thermal comfort, have you adjusted or done the following (adaptive behaviors)?". Figure 9.8 shows that the highest proportion (27.49%) of participants chose to adjust clothes to achieve thermal comfort. Next on the figure is hot/cold drink/food intake (20.70%), windows (14.06%), doors (10.74%), and reseat (10.66%). The other proportion of adaptive behaviors below 10% from high to low is curtains pulling, desk lamps, ceiling lights, air-conditioner, hot water bottles/ heaters use, and fans. As can be seen from the chart, occupants are more prone to adopt behaviors that have little impact on others.

Figure 9.10a shows proportion trends of different thermal adaptive behaviors in different operative temperatures. Although the proportion of adjusting clothes fluctuates, it keeps a high state. Occupants are more likely to intake hot/cold and drink/ food at cold temperatures. The behavior of window opening/close also keeps a high proportion. As shown in this figure, with the increase of operative temperature (Top), the proportion of people who chose to adjust their thermal comfort through

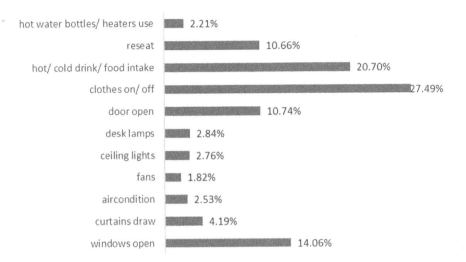

Fig. 9.8 Proportions of different thermal adaptive behaviors

beverages and food shows a decreasing trend to a certain extent, while the trend of adjusting their thermal comfort through window opening was on the contrary. There is a large difference in the proportion between the behaviors of windows and doors, two relatively similar behavioral characteristics. The reason may be that opening the window can keep the quiet and privacy of the classroom while adjusting the temperature.

Figure 9.9b shows a conclusion consistent with Fig. 9.9a. In addition, occupants opened/closed windows more often than to intake hot/cold drinkfood when they have a "warm" sensation.

9.3.4   Mean Thermal Sensation (MTS) and Predicted Mean Vote (PMV)

The relationship between predicted mean vote (PMV) and mean thermal sensation (MTS) in different phases is shown in Fig. 9.10. The average PMV and MTS values for each 1 °C interval were plotted. It can be seen that in NH phase 25.5 °C is a threshold after which MTS would be less than PMV. It indicates that participants are likely to feel warmer in this phase than expected. However, in the EH phase, MTS was always higher than PMV, indicating that the participants felt warmer than the PMV predictions under the same conditions. And in the LH phase, PMV has a greater slope and was much greater than MTS after the intersection in 23.6 °C, which indicates that in this phase participants felt colder than the PMV predictions. As the temperature rises, the actual thermal sensation of people is not so sensitive.

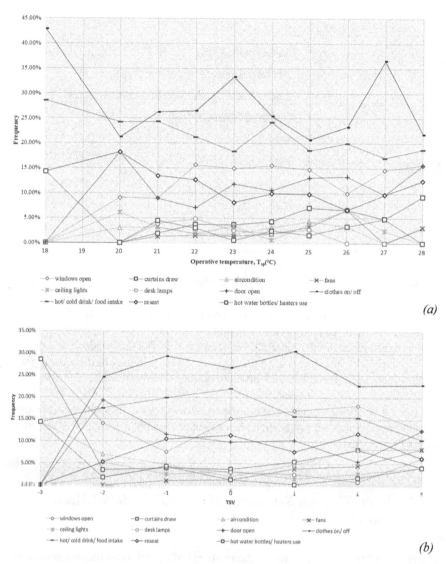

(a)

(b)

Fig. 9.9 Proportions trends of different thermal adaptive behaviors under (**a**) different operative temperatures and (**b**) different thermal sensation votes

9.3.5   Thermal Neutral Temperature

The thermal neutral temperatures of the three phases were shown in Fig. 9.11, taking gender into account. In the NH phase, the thermal neutral temperature was very close to the average indoor air temperature. In the EH phase, the thermal neutral temperatures increased and were a bit more than the indoor air temperature. But there a little dip when it turned into the LH phase, which was 0.4 °C lower than

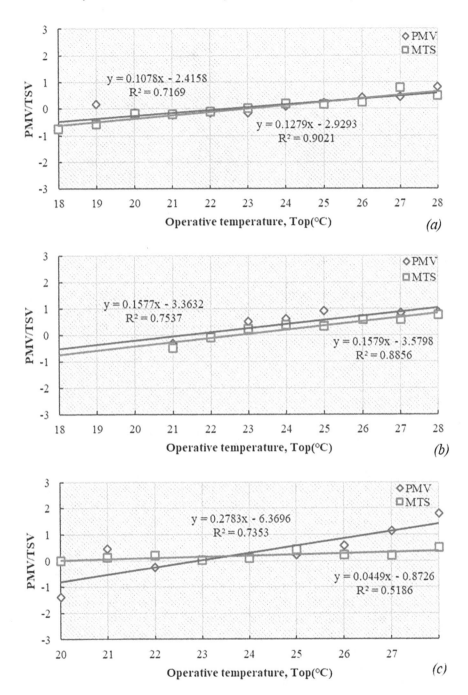

Fig. 9.10 Regression of MTS and PMV in (**a**) NH phase; (**b**) EH phase; (**c**) LH phase

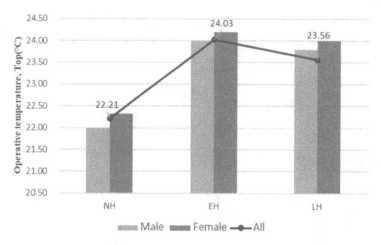

Fig. 9.11 Thermal neutral temperatures in different phases

indoor air temperature then. Although it shows the tendency to rise at the beginning and decline later, the thermal neutral temperature didn't change much overall, ranging from 2 °C. Besides, the neutral temperatures of females in all phases were slightly higher than that of males, between 0.2 and 0.5 °C.

9.4   Discussion

9.4.1   Frequency of Thermal Adaptive Behaviors

Thermal adaptive behaviors, as seen above, are one of the most important ways to keep comfortable in NV buildings. The thermal adaptation behaviors adopted by students can be roughly divided into two categories: one is spatial behavior, such as turning on and off windows, electric fans, and air conditioners; the other is self-centered behavior, such as adjusting clothes, ingesting food and beverages, changing seats, and turning on lamps and using warm water bags. As shown in Fig. 9.12a, most participants chose to take clothes on and off for thermal comfort among self-centered behavior. But this behavior does not show a strong functional correlation with the operative temperature ($R2 = 0.13 < 0.5$). While there was a stronger correlation between hot/cold drink/food intake and the operative temperature ($R2 = 0.70 > 0.5$). And Fig. 9.12b indicates that both curtains pulling and windows opening have a statistically significant correlation with $R2 = 0.85$ and 0.50, respectively. We can also see from the figures that participants were more likely to adopt spatial behaviors as the temperature increased. We can logically deduce the cause of this phenomenon is that, with the temperature rising, the regulating effect of an individual's behavior was limited, and it was difficult to reach the thermal comfort state. Therefore, people tended to adjust to the objective environment.

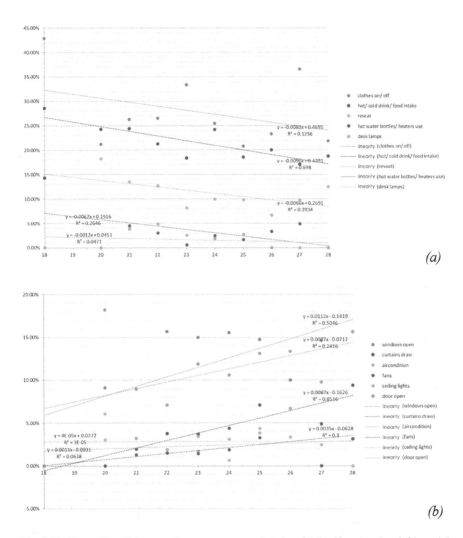

Fig. 9.12 Regression of the operative temperature and students' (**a**) self-centered and (**b**) spatial adaptive behaviors

9.4.2 Energy-Saving Potential

Indoor heating temperature requires careful consideration of individuals' adaptability to outdoor climates. Because the students had thermal adaptability to cold climates, they will be sensitive to the heating temperature in the classroom. However, students' thermal adaptability to the cold will be a weakness if they spend a long time at a high heating temperature.

When setting thermal neutral temperature as the indoor heating temperature, it is necessary to consider the health and wellbeing of students as well as energy consumption. The outdoor design temperature was set at −9 °C by the standard for Beijing's heating requirements, and the heating temperature is required to meet 18 °C in the interior. Assume that the thermal temperature was set to the current room temperature during the period of heating; the calculation formula is shown below:

$$N = \sum_{1}^{i} \left[\frac{K \times F \times (t'_{ni} - t_w) - K \times F \times (t_{ni} - t_w)}{K \times F \times (t_{ni} - t_w)} \times \frac{d_i}{D} \right] \times 100\%$$

N – the percentage of reduced energy.
t'_n – the temperature of the interior.
t_w – the design temperature of the exterior.
t_n – the thermal neutral temperature.
d_i – the number of days in each heating phase.
D – the total number of investigated days.
K – the heat transfer coefficient.
F – the size of external wall or window.
(Both K and F can be reduced in the formula.)

In EH phase: $N_{EH} = \left(\frac{20}{59} \times \frac{24.3 - 24.0}{24.0 - (-9)} \right) \times 100\% = 0.277\%$.

In LH phase: $N_{LH} = \left(\frac{19}{59} \times \frac{23.14 - 23.6}{23.6 - (-9)} \right) \times 100\% = -0.415\%$.

From the calculation of the above heating phase, in the EH phase, 0.277% of the heating energy can be saved, which means that the energy consumption has the potential to be decreased in the EH phase. While LH phase the result is −0.415%, because compared with current indoor air temperature, in this phase, students' thermal neutral temperature is higher.

In the NH phase, the student had already adjusted the low temperature. When the indoor temperature increased too quickly in EH, students can't adapt to the high temperature physiologically or psychologically. As a result, the overheating problem not only led to students' feel uncomfortable but also led to the waste of energy.

A period of heating allows the students to get used to the higher indoor environment, and they become more sensitive to the lower temperature. When the outdoor temperature gets lower in the LH phase, higher temperatures need to be provided to meet the students' thermal-neutral temperature.

Nowadays, building energy consumption occupies a large part of global energy consumption. When providing heating temperature in winter, the heating temperature should be taken into comprehensive consideration of the thermal neutral temperature. In the early heating period, the heating system can gradually improve temperature, and adjust the temperature according to outdoor temperature in middle and late heating periods, to achieve the goal of reducing energy consumption while satisfying the thermal comfort of the human body.

9.5 Conclusion

In this study, objective environmental parameter measurements and subjective questionnaires were simultaneously used to investigate the thermal comfort expectation and adaptive behavior characteristics of college students in Beijing. The main results of the study can be summarized as follows:

1. During the study time, the average indoor thermal atmosphere had a temperature of 22.87 °C and relative humidity of 31.09%. The temperature inside was between 20 and 25 °C for around 80% of the time. This meant that the temperature inside was a little cool.
2. The neutral operative temperature was determined by a linear regression study to be 23.1 °C. Although the tendency rose at the beginning and declined later, the thermal neutral temperature didn't change much overall, ranging from 2 °C. Besides, the neutral temperatures of females in all phases were slightly higher than that of males, between 0.2 and 0.5 °C.
3. Occupants are more prone to adopt behaviors that have little impact on others and more likely to adopt spatial behaviors as the temperature increased. The behaviors taken the most are taking clothes on and off, hot/cold drink/food intake, windows opening/close, door opening/close, and reseat. There were stronger correlations of "hot/cold drink/food intake," "curtains pulling," and "windows opening" with the operative temperature.
4. In winter, students' adaptation to the cold climate should be fully considered. A proper balance between comfort and energy effect should be considered. During space heating, unnecessary high thermal comfort zones should be avoided and appropriate temperature should be set. If the current indoor temperature is replaced by neutral temperature for heating, energy efficiency will be improved.

To summarize, this research looked into the thermal environment and proposed thermal comfort in China's coldest region. It has the potential to be useful. Building simulation may benefit from the functions of adaptive thermal behaviors. Furthermore, this field analysis would contribute to the creation of a more rigorous database for modeling optimum indoor thermal conditions.

Acknowledgments This work was funded by the Beijing Municipal Education Commission (Project number: KM201910016016) and the State Key Laboratory of Subtropical Building Science, South China University of Technology (Project number: 2019ZB09).

References

ASHRAE, ANSI/ASHRAE Standard 55—Thermal Environmental Conditions for Human Occupancy, American Society of Heating, Refrigerating and Air Conditioning Engineers, Atlanta, 2017.

Brager, G. S., & de Dear, R. J. (2000). A standard for natural ventilation. *ASHRAE Journal, 42*(10), 21–28.

CEN. EN 15251, Indoor Environmental Input Parameters for Design and Assessment of Energy Performance of Buildings Addressing Indoor Air Quality, Thermal Environment, Lighting and Acoustics, CEN (European Committee for Standardization), Brussels, 2007.

de Dear, R., Kim, J., Candido, C., & Deuble, M. (2015). Adaptive thermal comfort in Australian school classrooms. *Building and Environment, 43*(3), 383–398.

de Dear, R. J., & Brager, G. S. (1998). Developing an adaptive model of thermal comfort and preference. *ASHRAE Transactions, 104*(1), 145–167.

de Dear, R. J., & Brager, G. S. (2002, July). Thermal comfort in naturally ventilated buildings: Revisions to ASHRAE standard 55. *Energy and Buildings, Energy and Buildings, 34*(6), 549–561.

Eide, E. R., Showalter, M. H., & Goldhaber, D. D. (2010, February). The relation between children's health and academic achievement. *Children and Youth Services Review, 32*(2), 231–238.

Fanger, P. O. (1972). *Thermal comfort: Analysis and applications in environmental engineering.* McGraw-Hill.

GB, GB 50176: Code for Thermal Design of Civil Building, Ministry of Housing and Urban-Rural Development of the People's Republic of China, Beijing, 2016 (in Chinese).

Guerra-Santin, O., & Itard, L. (2010). Occupants' behaviour: Determinants and effects on residential heating consumption. *Building Research and Information, 38*(3), 318–338.

Haverinen-Shaughnessy, U., Shaughnessy, R. J., Colec, E. C., Toyinbo, O., & Moschandreas, D. J. (2015, November). An assessment of indoor environmental quality in schools and its association with health and performance. *Building and Environment, 93*(1), 35–40.

https://en.wikipedia.org/wiki/Beijing#Climate.

ISO, EN ISO 7730: 2005, Ergonomics of the Thermal Environment—analytical Determination and Interpretation of Thermal Comfort Using Calculation of the PMV and PPD Indices and Local Thermal Comfort Criteria, International Standardization Organization, Geneva, 2005.

Jindal, A. (Sep. 2018). Thermal comfort study in naturally ventilated school classrooms in composite climate of India. *Building and Environment, 142*, 34–46.

Jowkai, M., Rijal, H. B., Montazami, A., Brusey, J., & Temeljotov-Salaj, A. (2020, July). The influence of acclimatization, age and gender-related differences on thermal perception in university buildings: Case studies in Scotland and England. *Building and Environment, 179*(15).

Karjalainen, S. (2012, April). Thermal comfort and gender: A literature review. *Indoor Air, 22*(2), 96–109.

Liu, J, Luo, Q., & Cai, T. (2019, June). Students Responses to Thermal Environments in University Classrooms in Zunyi, China. *Materials Science and Engineering, Volume 592, International Conference on Manufacturing Technology, Materials and Chemical Engineering*, pp. 14–16.

López-Pérez, L. A., Flores-Prieto, J. J., & Ríos-Rojas, C. (2019). *Adaptive thermal comfort model for educational buildings in a hot-humid climate* (Vol. 150). Elsevier Ltd.

Mendell, M. J., Eliseeva, E. A., Davies, M. M., Spears, M., Lobscheid, A., Fisk, W. J., & Apte, M. G. (2013, December). Association of classroom ventilation with reduced illness absence: A prospective study in California elementary schools. *Indoor Air, 23*(6), 515–528.

Pérez-Lombard, L., Ortiz, J., & Pout, C. (2008). A review on buildings energy consumption information. *Energy and Buildings, 40*, 394–398.

Rijal, H. B., Humphreys, M. A., & Nicol, J. F. (2018 July). Development of a window opening algorithm based on adaptive thermal comfort to predict occupant behavior in Japanese dwellings. *Japan Architectural Review, 1*(3), 310–321.

Rupp, R. F., de Dear, R., & Ghisi, E. (2018). Field study of mixed-mode office buildings in Southern Brazil using an adaptive thermal comfort framework. *Energy and Buildings, 158*, 1475–1486.

Shrestha, M., Rijal, H. B., Kayo, G., & Shukuya, M. (2021, March). A field investigation on adaptive thermal comfort in school buildings in the temperate climatic region of Nepal. *Building and Environment, 190*, 107523.

terMors, S., Hensen, J. L. M., Loomans, M. G. L. C., & Boerstra, A. C. (2011, December). Adaptive thermal comfort in primary school classrooms: Creating and validating PMV-based comfort charts. *Building and Environment, 46*(12), 2454–2461.

Wang, D., Yan, J. J., Liu, F., Wang, Y., Xu, Y., & Liu, J. (2017a, April). Student responses to classroom thermal environments in rural primary and secondary schools in winter. *Building and Environment, 115*, 104–117.

Wang, Z., Ning, H., Zhang, X., & Ji, Y. (2017b). Human thermal adaptation based on university students in China's severe cold area. *Science and Technology for the Built Environment, 23*(3), 413–420. https://doi.org/10.1080/23744731.2016.1255495

Wang, Z., de Dear, R., Luo, M., Lin, B., He, Y., Ghahramani, A., & Zhu, Y. (2018, June). Individual difference in thermal comfort: A literature review. *Building and Environment, 138*(15), 181–193.

Yao, R. M., Liu, J., & Li, B. Z. (2010). Occupants' adaptive responses and perception of thermal environment in naturally conditioned university classrooms. *Applied Energy, 87*(3), 15–22.

Zomorodian, Z. S., Tahsildoost, M., & Hafezi, M. (2016, June). Thermal comfort in educational buildings: A review article. *Renewable and Sustainable Energy Reviews, Elsevier Ltd, 59*, 895–906.

Chapter 10
User Satisfaction Feedback and Environmental Comfort Evaluation of Sustainable Building: Take Huahui Research and Design Center as an Example

Pengfei Wang, Junjie Li, and Yichun Jin

10.1 Preface

The green building, also known as ecological building or sustainable building, is an environmental protection system that follows the basic principles of sustainable development such as "protecting the earth's environment, saving resources, and ensuring the quality of human settlements" (China Academy of Building Research GB/T50378-2006, 2006). In recent years, in the context of the energy crisis, green buildings have become an important branch of the rapid development of the construction industry due to their energy-saving and sustainable advantages. Since the publication of GB/T50378-2006 "Assessment Standard for Green Building" in 2006, the number of green buildings in China has been increasing year by year. This paper takes a public building in Shaoxing City, Zhejiang Province, that has obtained the 3-star levels of Green Building as an example to conduct research and analysis, to master the application and operating effects of its green building technology, and to gain an in-depth understanding of the development status of green buildings in China.

P. Wang
School of Architecture, Southeast University, Nanjing, China

School of Architecture and Design, Beijing Jiaotong University, Beijing, China

J. Li (✉) · Y. Jin
School of Architecture and Design, Beijing Jiaotong University, Beijing, China
e-mail: lijunjie@bjtu.edu.cn

10.2   Basic Situation of the Survey Object

10.2.1   Basic Overview of the Project

Shaoxing City, Zhejiang Province, is located in the hot-summer and cold-winter zone, with rain and heat at the same time and abundant sunshine. Huahui Research and Design Center is located on the Jiefang Avenue in Shaoxing (Fig. 10.1). It was designed and constructed from 2012 to 2016. The Miaoqian River and Meishan River flow through the south and east sides of the building, respectively.

10.2.2   Project Composition and Layout

The project is a scientific research office building, covering an area of 3131.17 square meters and a total construction area of 39436.82 square meters. It adopts a frame shear wall structure, with 23 floors above ground, 2 floors underground, and low podiums on the 4th floor and below. The project has two entrances to the east and north. In order to create a shared communication space, the building starts from the fifth floor, with roof terraces on the east side every other floor. The podium part of the building is a rectangular plan, and the high-rise part is a square layout. With

Fig. 10.1 Huahui Research and Design Center

the traffic in the center of the plan as the core, independent and centralized office spaces are arranged around.

10.2.3 Analysis of Green Building Technology

The main technical measures applied in this project will be analyzed according to the principle of "a green four sections."

1. Land-saving and outdoor environment: The project is a high-rise building. The underground space is reasonably developed for parking and equipment rooms, and the space utilization rate is high. There is no fence in the site, and the green area rate is 14.9%, but they are all open green areas, which can provide leisure places for surrounding residents.
2. Water-saving and water resource utilization: The project adopts a rainwater recycling system to fully collect roof rainwater and road drainage and set up a rainwater reservoir in conjunction with the Miaoqian River on the south side of the site, which is mainly used for landscape water and green irrigation. Water-saving appliances are used for water supply appliances in the building, and the water-use efficiency level reaches level 1.
3. Energy conservation and energy utilization: 1. Air-conditioning system – The project makes full use of the surrounding river water resources. The water source heat pump air-conditioning system is adopted for the 19th floor and below, and the VRV air-conditioning system is adopted for the water lift problem above the 19th floor. Water cold and heat storage system responds to peak and valley electricity price differences. 2. Lighting system – Use voice-activated switches in the stairwell, and use a combination of general lighting and local lighting in the office area. Design light guide lighting systems in basements and restaurants to enhance natural lighting to a certain extent. 3. Solar photovoltaic system – Install polysilicon solar photovoltaic panels on the roof of the main building and podium, with a total installation area of about 1190 square meters. The power generation adopts spontaneous generation and self-use, and the residual power is connected to the Internet (Huahui Research and Design Center, n.d.).
4. Material-saving and material utilization: 1. Foundation pit support – Under the premise that the excavation depth remains unchanged, the two supports of the two-story basement are changed to one support, and gussets are locally added at the weak links of the structure to avoid large-scale supporting enclosures. 2. Reduce pile foundations – Use the favorable buoyancy effect of the groundwater level to test piles under controlled experimental conditions to achieve a reasonable balance between the excavation of the ground bearing capacity and the reduction of the number of pile foundations. The optimized pile foundation cost can be reduced 15%.
5. Environmental protection: Use watering, covering, and other noise and dust reduction measures during the construction process. The project design vertical

greening, roof greening and vegetable gardens, plant trees on both sides of the road, etc., which not only improves the indoor noise environment of the building but also improves the environmental quality of the site.

10.3 Research Methods and Objectives

For the physical environment test of the project, this survey will be carried out in the transitional month, which can be compared with the winter and summer. This paper draws on relevant literature and uses the two most important indicators of existing evaluation standards, building energy consumption and indoor environmental quality as the two important first-level indicators for investigation and analysis (Jihong Shang, 2018). The indoor environmental quality is based on the literature. It is further streamlined based on the actual conditions of the investigation and only contains three secondary indicators of temperature, indoor air quality, and illuminance.

10.3.1 Site Visit and Survey

Through site visits and surveys, we can understand the basic information of the building, the functional layout of the building's internal space, the current status of use, the specific aspects of the application of green technology and the completion status, etc. Ask the relevant technical personnel about the energy consumption of the building operation and gain a preliminary understanding of the indoor environmental quality and personnel satisfaction through exchanges and visits with office personnel.

10.3.2 Use Tools to Measure

In the indoor environmental quality measurement, because the two underground floors are parking spaces, no measurement is made. The above-ground building is divided into one to four-story podium and the five- to twenty-three-story main building.

Temperature and illuminance tests were conducted using infrared radiation thermometers and handheld illuminance meters. The first, fourth, thirteen, and twenty-second floors were selected. And they were arranged at three-hour intervals at 17:00, 20:00 on March 22, 2019, and 09:00, 15:00, and 18:00 on March 23, 2019. Because the building plan is relatively regular, the measuring points are evenly distributed according to the structural column network. Some rooms have no measuring points because they are not accessible. Specific measuring points distribution are shown in the figure below (Fig. 10.2).

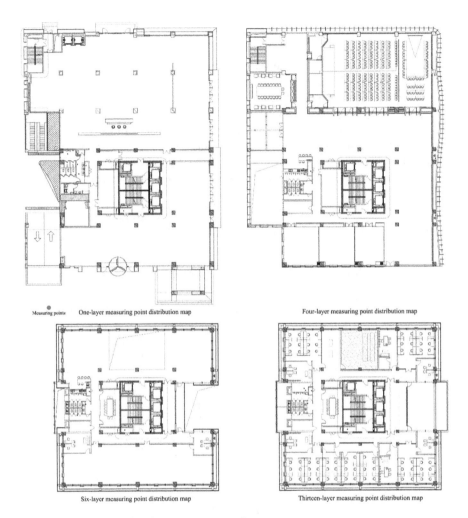

Fig. 10.2 Distribution of project measurement points

Indoor air quality measurement includes five items: carbon dioxide, wind speed, PM2.5, PM10, and formaldehyde. The instruments used are carbon dioxide recorder, handheld anemometer, and electrochemical sensor. Measurement selections are typical indoor open office areas of the first, fourth, fifth, sixth, thirteenth, and twenty-second floors.

We adopt the method of continuous monitoring for carbon dioxide, and the test time is from March 22 to March 24, 2019. In addition to the above-mentioned measuring points for wind speed measurement, an outdoor and five-story terrace space are added.

10.3.3 Questionnaire Survey

Use questionnaire surveys to understand the satisfaction of the staff in the building with the implementation of the green construction technical measures. The level of satisfaction can be used as a reference for the actual application of the technology and timely feedback on problems in the building. The content of the questionnaire was adjusted in accordance with the abovementioned principle of "a green four sections," which mainly included five aspects: land-saving, water-saving, material-saving, indoor environmental quality, and operation management. Respondents determined their satisfaction in related aspects according to the scale (Fig. 10.3).

10.4 Analysis of Survey Results

10.4.1 Building Operation Energy Consumption

According to relevant technical data, the annual heating energy consumption of the project is 48.51 kW·h/m², the annual air conditioning energy consumption is 23.05 kW·h/m², the annual total energy consumption is 71.56 kW·h/m², and the energy-saving rate is 52.79%, which meets the requirement of 50% (implementation standard) in the "Design Standard for Energy Efficiency of Public Buildings" in Zhejiang Province. The figure below shows the comparison of the annual calculation load of heating and air-conditioning between the reference building and the actual building (Fig. 10.4). The project load is reduced by 6.64%. It can be seen that the green building has a significant effect on saving energy consumption, but there is still much room for development.

10.4.2 Indoor Environmental Quality

1. Temperature

Select the temperature data of the measurement points on each floor at 15:00 on March 23, and draw the temperature analysis diagram and the measurement point data cloud diagram (Fig. 10.5). It can be seen that the temperature of each floor is distributed between 15 and 20 °C, and the temperature fluctuation range is about

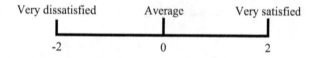

Fig. 10.3 Staff satisfaction scale

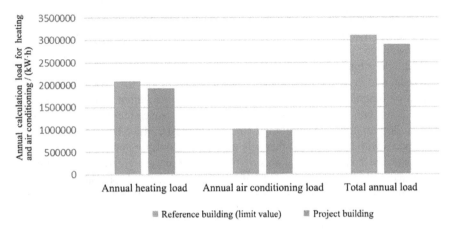

Fig. 10.4 Comparison of annual calculation load of heating and air-conditioning

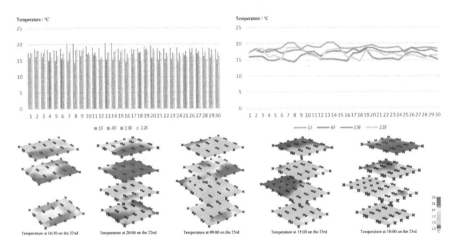

Fig. 10.5 Analysis of temperature measurement data

3 °C at each measuring point on the same floor during the same time period. The temperature difference between the edge of the building plane and the center is large, and the indoor environment temperature is not necessarily related to the number of floors.

2. Illumination

In order to avoid glare in the traditionally designed office area, thick blackout curtains are usually hung, which leads to the dilemma of brightly lit indoors during the day. For this purpose, the project innovatively designed an external wall window system. The upper part of the window is equipped with a horizontal shading light guide plate, which diffuses the soft light into the depths of the room; the central panoramic window makes the city view clearer, and the shading curtain inside the window in summer can effectively reduce heat radiation; low-position mid-hung window design solves the contradiction between shading and ventilation, so that natural wind can pass through the designer's workstation.

Select the illuminance data of the measurement points on each floor at 15:00 on March 23, and draw the illuminance analysis diagram and the measurement point data cloud diagram (Fig. 10.6). It can be seen that the indoor illuminance of the project is not enough. Many measuring points do not reach the design illuminance value of 300 lux for office buildings. Because domestic office buildings are mostly open spaces with large depths, they rely on side windows for natural lighting. At the same time, the measurement points on the same floor vary greatly, and the indoor illuminance uniformity is poor. The light near the window is good and even reaches about 2500lux, which is easy to produce the problem of glare. While the location far away from the window has serious illuminance attenuation, the light is relatively dim, which needs to be supplemented by artificial lighting. The uneven indoor light environment can easily cause visual fatigue of office workers (Bangwei Wan, 2017).

3. Indoor air quality

1. Carbon dioxide volume fraction

Statistics on the openable area of the external windows of the building's natural ventilated rooms show that the openable area of a single natural ventilated room is greater than 30% of the external window area, which can effectively improve the efficiency of natural ventilation. At the same time, the design ensures that the air age of the main functional rooms is less than 1800s under natural ventilation conditions to meet the requirements of green buildings.

Taking the carbon dioxide data record from 9:00 to 18:00 on March 23, 2019, during the survey time as an example analysis (Fig. 10.7), it can be seen that the indoor carbon dioxide concentration of the building has no obvious

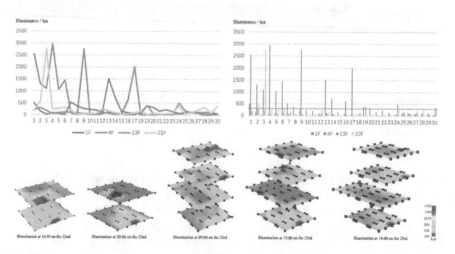

Fig. 10.6 Illumination measurement data analysis

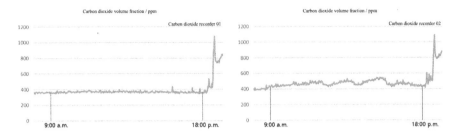

Fig. 10.7 Line graph of carbon dioxide measurement data

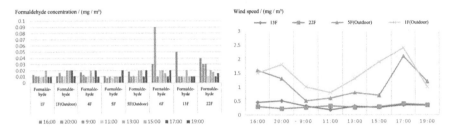

Fig. 10.8 Analysis of formaldehyde concentration and wind speed measurement data

change during the working hours. It is basically stable at about 400–450 ppm, much lower than 1000 ppm. Indoor ventilation is good, but the volume fraction varies slightly between different floors. Good ventilation has a certain relationship with the "light guide, shading and ventilation composite passive window-wall system" adopted by the project.

2. Wind speed: Indoor wind speed is obviously lower than outdoor, the wind speed in the outdoor 1.5 m height area is less than 3 m/s, and the indoor wind speed value is generally below 0.5 m/s, which is suitable (Fig. 10.8).

3. The average concentrations of PM2.5 and PM10 are around 30 μg/m³ and 40 μg/m³, respectively, which have reached the corresponding national current first-level standards (Fig. 10.9). There is no obvious difference in the concentration of the two in each floor, and the numerical value has no obvious change trend over time. The formaldehyde concentration is less than 0.03 mg/m³ during most of the measurement time, which meets the requirements of relevant standards.

10.4.3 The Satisfaction Questionnaire Survey Results

Post-occupancy evaluations can be seen as useful tools for obtaining feedback on how buildings perform and how they are experienced by their inhabitants after they have been occupied (Dursun & Ahsen, 2008). In this survey, a total of 53 questionnaires were issued, and 53 valid questionnaires were actually collected. The author

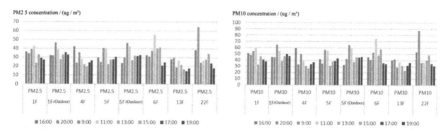

Fig. 10.9 PM2.5 and PM10 measurement data analysis

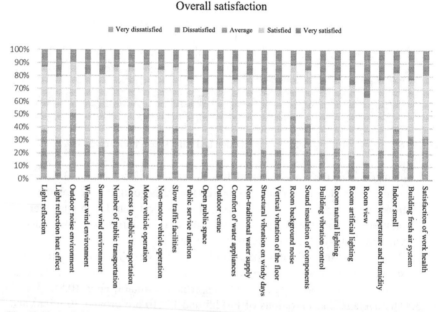

Fig. 10.10 Individual satisfaction of questionnaire analysis

organizes and analyzes the questionnaire and draws the survey research (Fig. 10.10). It can be concluded that about 47% of the subjects who are satisfied with green building technical measures. Among them, the satisfaction rate for material saving and material utilization is the highest, while the satisfaction rate for land saving and outdoor environment is the lowest (Fig. 10.11).

Selecting a single item for analysis shows that the interviewees are most dissatisfied with the outdoor noise environment (Fig. 10.12) and most satisfied with the visual field of the room they are in (Fig. 10.13). The two are actually in a mutually restrictive relationship. Because the east side of the project is the Jiefang Avenue, traffic is the main source of noise. The exterior walls of the building use components such as aerated blocks, sound reinforcement panels, and sound insulation panels to weaken the impact of the building. The office space is arranged far away from the road, so that the indoor noise level reaches the average value of the low and high standard limit, but

Overall satisfaction

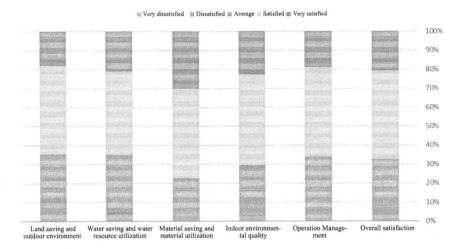

Fig. 10.11 Comprehensive satisfaction analysis

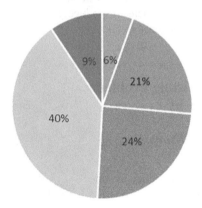

■ Very dissatisfied ■ Dissatisfied ■ Average ■ Satisfied ■ Very satisfied

Fig. 10.12 Satisfaction of outdoor noise environment

does not reach the high standard limit. Adjacent to the city's arterial roads, the design of horizontal windows on the east, south, and north sides of the main building strives for maximum lighting, ventilation, and landscape for the interior. But while obtaining a good view, it will also reduce the blocking effect on environmental noise.

In addition, the light reflection environment satisfaction performance of the building during the day is acceptable, but the survey found that there is still strong glare at the window-side workstations. The project designed climbing plants on the west side of the main building as shading materials. Setting louvers outside the equipment platform can not only reduce the air temperature around the equipment,

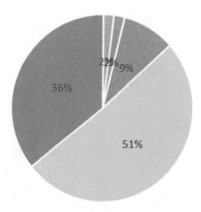

■ Very dissatisfied ■ Dissatisfied ■ Average ■ Satisfied ■ Very satisfied

Fig. 10.13 Satisfaction of the room's vision

but also make the facade more integrated. In addition, there is a movable external sunshade outside the west window, which can be opened according to the indoor light demand and comfort. These measures reduce direct sunlight to a certain extent and achieve the purpose of low-cost green.

10.5 Analysis of Existing Problems and Suggestions for Improvement

10.5.1 Wall Sound Absorption and Sound Insulation

In the design stage, in order to reduce the impact of traffic noise on the office area, the office was arranged as far away as possible from the Jiefang Avenue, trees were planted along the road, and vertical greening was planted around each floor, and glass baffles were installed to reduce noise. But the building is close to the main road of the city, walls should pay more attention to the improvement of sound absorption and sound insulation performance. The existing method of combining aerated blocks with facade components and vertical greening has not yet reached the user's requirements in the satisfaction test.

10.5.2 Improvement of Shading Conditions

Because office buildings mostly adopt the layout characteristics of large space and deep depth, and emphasize the permeability of the facade, the indoor lighting level and uniformity need to be further improved. At the same time, attention should be paid to avoiding glare at the window position. Currently, manual adjustment of the inner shade curtain is adopted, which is relatively low in reliability. In order to avoid direct sunlight, users often use artificial lighting after pulling down the shade curtain, which is not conducive to building energy saving.

10.5.3   Standardization of the Construction Process

The survey found that many skilled workers in the construction process of the project were not of professional origin, which resulted in the insufficient use of construction technology and green technology, insufficient management, and rework. That affected the energy-saving effect and technical completion effect. It is necessary to strengthen the concept of digital construction which needs to be realized throughout the process.

10.5.4   Focus on Operation and Management

In order to achieve the goal of the 3-star levels in operating, the project needs to strengthen operation and maintenance in the long-term management and use process. There is still a lot of room for development of green and energy-saving projects. The project can strengthen the publicity and education of green property management (Fangyan Li, 2013), and enhance the energy-saving awareness and behavior of users and managers.

10.6   Conclusion

As a design and construction work completed by a design institute itself, the Huahui Research and Design Center has a good demonstration. The use of various green building technologies is reasonable and effective. Technical measures such as vertical greening, passive window-wall systems, and water source heat pumps are unique and innovative. It provides a reference for the construction of green buildings under the same geographical environmental conditions and plays a good role in promoting the development of green buildings in Shaoxing area and even the entire Zhejiang province.

References

Bangwei Wan. (2017). Field survey and analysis of indoor light environment of office buildings in Nanjing [J]. *Architectural Journal, 4,* 118–123.

China Academy of Building Research GB/T50378-2006. (2006). *Assessment standard for green building [S].* China Architecture & Building Press.

Dursun, P., & Ahsen, Z. S. O. Y. (2008). How can architects learn from their own experiences? [J]. *A|Z, 5*(2), 82–95.

Fangyan Li. (2013). Analysis for implementation of Green Building Technologies in office building [J]. *Construction Technology, 42,* 12–13.

Huahui Research and Design Center Green Building Operation Mark Application Form.

Jihong Shang. (2018). Investigation and evaluation of operation status of green office buildings in Changping District, Beijing [J]. *Heating Ventilating & Air Conditioning, 48,* 51–52.

Chapter 11
Using Web Data Scraping to Reveal the Relationship Between AI Product and Room Layout

Xiao Guo, Zhenjiang Shen, Xiao Teng, and Yong Lin

11.1 Introduction

It is common knowledge that the development of technologies leads to many changes in our lives. Building, as known as the most common thing people live in every day, has also been improving. Among many technologies, artificial intelligence (AI) is showing rapid growth. It can be combined with smart building technology to become an innovative tool. Some companies have already manufactured some smart home AI products, such as Apple, Google, Xiaomi and Amazon, etc. AI products in this article are defined as industrial products equipped with artificial intelligence systems, which can realize functions such as voice recognition, image recognition, autonomous decision-making, and so on, to help users live more efficiently and comfortably. Sensors are also very important for AI products, so in this research, we also scraped data of some sensors. Since most of the smart building products are smart home products, AI products selected in this study are all smart home products. People buy these AI products and place them in their house to improve security, entertainment, and building environment.

Many previous studies have established that AI technology can make a building smarter in many ways. Some researchers try to use AI technology to manage devices

X. Guo · Y. Lin
School of Natural Science and Technology, Environmental Design Division, Kanazawa University, Kanazawa City, Japan

Joint International Laboratory of Spatial Planning and Sustainable Development (FZUKU-LAB SPSD), Fuzhou, China

Z. Shen (✉) · X. Teng
School of Natural Science and Technology, Environmental Design Division, Kanazawa University, Kanazawa City, Japan
e-mail: shenzhe@se.kanazawa-u.ac.jp

© The Author(s), under exclusive license to Springer Nature Switzerland AG 2021
S. S. Y. Lau et al. (eds.), *Design and Technological Applications in Sustainable Architecture*, Strategies for Sustainability, https://doi.org/10.1007/978-3-030-80034-5_11

in a smart building. A system called Rudas comprised of few embedded sensors and networking devices in the building and implanted a fuzzy logic artificial intelligence to adjust the intensity of light and air condition in a room (Crisnapati et al., 2016). Intelligent control in the smart house can be realized by analyzing the data in a sensor network and the user's previous behavior of operation to the household appliances, without the user's intervention (Shao, 2015). Some researchers try to use AI technology to improve security in a smart building. Machine learning techniques can be used to detecting anomalous events or actions in smart environment datasets for enhancing living experience (Jakkula & Cook, 2011). A wearable motion sensing device and its corresponding 3D gesture recognition algorithm can implement a convenient automated household appliance control system (Hsu et al., 2017). Some researchers try to use AI technology to achieve intelligent interaction in a smart building. An intelligent robot companion is capable of adapting the ambient conditions of a smart home environment in accordance with the user behavior and facilitate voice interaction (Senarathna et al., 2018). Smart building devices can be controlled and manage by Xenia, a secure smart home platform through text or voice chatting in natural English language (El Mougy et al., 2017). Some researchers attempted to explore systematically how and why people use smart home technologies and what impact on different aspects of domestic life (Wilson et al., 2015). Some researchers attempt to reveal the relationship between AI technology and smart building. A review of AI product and smart building shows that AI technology helps smart building in device management, energy management, healthcare, intelligent interaction, security, entertainment systems, and personal robots by utilizing activity recognition, data processing, decision-making, image recognition, prediction making, and voice recognition (Guo et al., 2019). However, we still do not know where to install those AI products that can make full use of them. There is a lack of a connection between the architect (design room layout) and the product designer (arrange AI product) which makes an absence of corporation.

This research aims to figure out the relationship between AI products and building layout. To address this objective, we need to answer two research questions: Where can we get the latest and reliable information? How to evaluate the performance and location of AI products? The framework of this study is shown in Fig. 11.1. We conducted a web data scraping via Gooseeker in Amazon.com with AI products' review title, review content, review date, username, and rating. Ten different functions of AI products are selected, namely, smart camera, temperature/humidity sensor, smart speaker, thermostat, door/window sensor, motion sensor, smoke/carbon detector, smart light, smart hub, and smart screen. Because most of the AI products are smart home products, the room types in this research are also common residential room types. Then, we used the name of rooms as keywords to filter the results, namely, kitchen, dining room, living room, bathroom, hallway/foyer/entrance, and bedroom. After we got the result of data collection, we used quantitative approaches to analyze those data. Finally, we can learn where people put AI products most and where AI products have the most satisfaction.

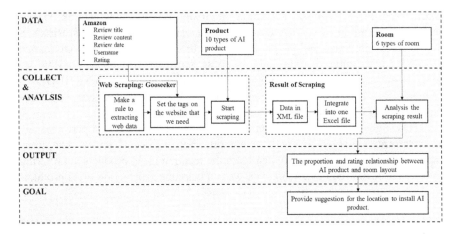

Fig. 11.1 Research framework

11.2 Method

In order to discover the relationship between AI products and room layout, a web scraping was conducted. This section describes the selected products and data, instrument, and procedure of the research.

11.2.1 Selected Products and Data

Among many smart building-related products, 30 products were selected in this research. In order to facilitate web scraping, all the products are from Amazon.com, since it has many users and reviews, and most of the reviews are in English. They are in ten different functions, namely, smart camera, temperature/humidity sensor, smart speaker, thermostat, door/window sensor, motion sensor, smoke/carbon detector, smart light, smart hub, and smart screen. Each function has three products as representatives. For smart camera, we chose Canary Flex, Wyze Cam 1080p HD, and Nest Cam 1080p HD. For temperature/humidity sensor, we chose SensorPush Wireless Thermometer/Hygrometer, Nest Temperature Sensor, and Govee Wireless Thermometer Hygrometer. For smart speaker, we chose Echo Dot, Amazon Tap, and Sonos One. For thermostat, we chose ecobee4 Smart Thermostat, Nest Learning, and Honeywell Home RTH9585WF1004 Wi-Fi Smart Color Thermostat. For door/window sensor, we chose Ring Alarm Contact Sensor, Samsung Smart Things GP-U999SJVLAAA Door and Window Multipurpose Sensor, and GE Personal Security Window Door Alarm. For motion sensor, we chose Kangaroo Home Security Motion Sensor, Philips Hue Indoor Motion Sensor, and Samsung SmartThings GP-U999SJVLBAA Magnetic Motion Sensor. For smoke/carbon detector, we chose Alexa-Enabled Smoke Detector and Carbon Monoxide Detector

Alarm, Nest Protect Smoke + Carbon Monoxide Alarm, and Onelink Smoke Detector and Carbon Monoxide Detector. For smart light, we chose BERENNIS Smart Light Bulb, Smart WiFi Light Bulb, and LIFX 1100-Lumen, 11W Dimmable A19 LED Light Bulb. For smart hub, we chose Samsung GP-U999SJVLGDA 3rd Generation Smart Things Hub, Philips Hue Smart Hub, and Wink WNKHUB-2US 2 Smart home hub. For smart screen, we chose Echo Spot, Nucleus Anywhere Intercom, and Echo Show.

To investigate where we should put AI products can have the best performance, out of information about products, review title, review content, review date, username, and rate were scraped. We can learn the location of AI products via filtering keywords in review content. Moreover, we can learn the satisfaction of AI products in a specific room by the rating.

11.2.2 Instrument

Since we need to process a huge amount of products reviews data, the method of manually collecting reviews is not applicable. We decided to use web data scraping software instead of manual collection of review information. Web scraping is a technique for extracting data from websites. It converts unstructured data into structured data that can be stored in a local computer or database. There are many free web scraping softwares, such as Beautiful Soup, Octoparse, and ParseHub. A web data scraping tool named Gooseeker was used to collect product data in this research. This software can automatically extract data from the web without coding and turn web pages into structured data within clicks. We can open a website in the built-in browser and start a scraping task by clicking and dragging.

11.2.3 Procedure

After selected products, open an amazon customer reviews page of one product in the built-in browser of Gooseeker. Then, we defined a rule to scrap review title, review content, review date, username, and rate of one product into XML files. Also, the rule of automatically loads more pages when scrolling down is necessary. The same rule was applied for all 30 products. Then convert XML files to one Excel file in Gooseeker. After that, we used Data > Filter > Text filters > Contains to search keywords in review content and record reviews number and average rate. There may be cases where there are multiple names for a room, but this research only involves the issue of proportions, so it has little effect on the results. Those keywords are the most common name of different types of rooms, namely, kitchen, dining room, living room, bathroom, hallway/foyer/entrance, and bedroom. Specifically, use Filter > Advanced to search multi-keywords. Approximately 89% of reviews do not mention specific room location, so we abandoned these reviews.

Then we summarized all the 30 products data into one Excel file and analyzed the proportion and average rate of each room. For the most placed positions, we manually read product reviews and qualitatively discussed why people chose to put this product in this room.

11.3   Result and Discussion

The result of 30 Amazon.com AI products data extraction is shown in Table 11.1. The data analysis results of the web scraping can be found in Table 11.2 and Table 11.3.

As shown in Table 11.2, smart camera is most often placed in the living room and most rarely in the dining room. Temperature/humidity sensor is most often placed in the bedroom and most rarely in the dining room. Smart speaker is most often placed in the kitchen and most rarely in the hallway. Thermostat is most often placed in the bedroom and most rarely in the hallway. Door/window sensor is most often placed in the hallway and most rarely in the dining room. Motion sensor is most often placed in the hallway and most rarely in the dining room. Smoke/carbon detector is most often placed in the bedroom and most rarely in the dining room. Smart light is most often placed in the bedroom and most rarely in the hallway. Smart hub is most often placed in the bedroom and most rarely in the dining room. Smart screen is most often placed in the bedroom and most rarely in the dining room.

As shown in Table 11.3, smart camera has the highest rating in the dining room and has the lowest rating in the living room and bedroom. Temperature/humidity sensor has the highest rating in the hallway and has the lowest rating in the dining room. Smart speaker has the highest rating in the bathroom and has the lowest rating in the hallway. Thermostat has the highest rating in the hallway and has the lowest rating in the kitchen. Door/window sensor has the highest rating in the bathroom and has the lowest rating in the living room. Motion sensor has the highest rating in the bathroom and has the lowest rating in the living room. Smoke/carbon detector has the highest rating in the bathroom and has the lowest rating in the dining room. Smart light has the highest rating in the hallway and has the lowest rating in the dining room. Smart hub has the highest rating in the living room and has the lowest rating in the dining room. Smart screen has the highest rating in the bedroom and has the lowest rating in the hallway.

We can see from Fig. 11.2 that sometimes the products in the most frequently placed rooms get the lowest ratings. For example, smart camera often placed in the living room and only get a 4.0 average rate. Since the most placed position has the most reviews, we manually read those reviews and got the reasons why people put the product in this room and some problems or suggestions.

The reasons why people most often place smart camera in the living room are as follows:

1. To monitor the situation of home and the safety of valuables.

Table 11.1 Result of Amazon.com AI product data extraction

Category	Product	Kitchen Num	Proportion of Num	Kitchen Rate	Dining Room Num	Proportion of Num	Dining Room Rate	Living Room Num	Proportion of Num	Living Room Rate	Bathroom Num	Proportion of Num	Bathroom Rate	Foyer/Hallway/Entrance Num	Proportion of Num	Foyer/Hallway/Entrance Rate	Bedroom Num	Proportion of Num	Bedroom Rate
Camera	1	14	25%	3.5	1	2%	5	22	39%	3.82	1	2%	5	9	16%	4.11	9	16%	4
	2	23	25%	4.48	2	2%	4.5	27	29%	3.85	6	7%	4.5	14	15%	4.64	20	22%	3.85
	3	11	20%	4.36	2	4%	4.5	23	43%	4.34	2	4%	5	6	11%	4.17	10	19%	4.3
Total			23%	4.17		3%	4.60		37%	4.00		4%	4.57		14%	4.43		19%	4.00
Temperature/Humidity Sensor	1	12	30%	4.42	0	0%	/	6	15%	4.33	2	5%	5	2	5%	5	18	45%	4.61
	2	7	9%	4	1	1%	4	13	16%	4.31	1	1%	4	8	10%	4.63	51	63%	4.59
	3	0	0%	/	0	0%	/	6	75%	4.5	0	0%	/	0	0%	/	2	25%	5
Total			13%	4.27		0%	4		35%	4.36		2%	4.67		5%	4.70		44%	4.61
Smart Speaker	1	105	38%	4.47	5	2%	4.2	50	18%	4.5	28	10%	4.64	5	2%	3.4	82	30%	4.44
	2	59	43%	4.51	4	3%	4.5	20	14%	4.4	17	12%	4.82	1	1%	5	37	27%	4.49
	3	10	23%	3.5	2	5%	5	16	37%	3.63	4	9%	5	0	0%	/	11	26%	4.64
Total			35%	4.43		3%	4.45		23%	4.31		11%	4.73		1%	3.67		27%	4.47
Thermostat	1	20	14%	3.75	8	5%	4	29	20%	3.9	2	1%	3	20	14%	4.15	67	46%	4.22
	2	3	7%	3	4	9%	4	8	18%	4.5	2	4%	5	9	20%	4.56	19	42%	3.84
	3	1	4%	5	2	7%	5	4	15%	4	0	0%	/	6	22%	3.83	14	52%	4
Total			8%	3.71		7%	4.14		17%	4.03		2%	4.00		19%	4.20		47%	4.12
Door/Window Sensor	1	0	0%	/	0	0%	/	0	0%	/	0	0%	/	3	75%	4	1	25%	5
	2	2	40%	3	0	0%	/	2	40%	2	0	0%	/	1	20%	5	0	0%	/
	3	5	9%	5	0	0%	/	3	5%	4	10	17%	4.9	5	9%	3.6	35	60%	4.66
Total			16%	4.43		0%	0		15%	3.20		6%	4.9		35%	3.89		28%	4.669444

Product		Kitchen Num	Kitchen Proportion of Num	Kitchen Rate	Dining Room Num	Dining Room Proportion of Num	Dining Room Rate	Living Room Num	Living Room Proportion of Num	Living Room Rate	Bathroom Num	Bathroom Proportion of Num	Bathroom Rate	Foyer/Hallway/Entrance Num	Foyer/Hallway/Entrance Proportion of Num	Foyer/Hallway/Entrance Rate	Bedroom Num	Bedroom Proportion of Num	Bedroom Rate
Motion Sensor	1	2	8%	5	0	0%	/	3	13%	4	8	33%	4.38	6	25%	4	5	21%	4
	2	24	23%	4.17	3	3%	4	8	8%	3.75	32	30%	4.41	30	28%	4.47	9	8%	4.11
	3	5	33%	3.6	0	0%	/	0	0%	/	3	20%	5	6	40%	4	1	7%	5
Total			21%	4.13		1%	4		7%	3.82		28%	4.45		31%	4.34		12%	4.13
Smoke/	1	13	27%	4.23	1	2%	1	2	4%	4	0	0%	/	13	27%	3.77	20	41%	3.35
Carbon	2	40	22%	4.5	1	1%	5	9	5%	3.56	12	7%	4.5	66	37%	4.65	51	28%	4.04
Detector	3	3	33%	3.33	0	0%	/	1	11%	1	0	0%	/	2	22%	4	3	33%	3.33
Total			27%	4.37		1%	3.00		7%	3.42		2%	4.5		29%	4.49		34%	3.82
Smart Light	1	8	9%	4.48	8	9%	4.63	26	29%	4.73	3	3%	5	3	3%	4.67	42	47%	4.5
	2	13	11%	4.77	6	5%	4.33	42	36%	4.5	0	0%	/	4	3%	5	52	44%	4.4
	3	5	13%	5	2	5%	3.5	13	34%	4	3	8%	4	0	0%	/	15	39%	4.2
Total			11%	4.73		6%	4.38		33%	4.49		4%	4.5		2%	4.86		44%	4.41
SmartHub	1	1	10%	2	0	0%	/	1	10%	5	2	20%	4	2	20%	2.5	4	40%	4.5
	2	6	23%	3.83	1	4%	3	5	19%	3.4	2	8%	2	2	8%	4.5	10	38%	3.4
	3	6	35%	3.83	0	0%	/	5	29%	4	3	18%	4	0	0%	/	3	18%	3.67
Total			23%	3.69		1%	3.00		20%	3.82		15%	3.43		9%	3.50		32%	3.71
Smart Screen	1	54	24%	4.31	1	0%	4	32	14%	4.34	8	4%	4	0	0%	/	133	58%	4.27
	2	5	36%	4.2	0	0%	/	4	29%	5	0	0%	/	1	7%	5	4	29%	4.75
	3	149	40%	3.99	2	1%	3.5	49	13%	3.8	20	5%	4	4	1%	2	152	40%	4.04
Total			33%	4.08		0%	3.666667		19%	4.06		3%	4		3%	2.60		42%	4.16

Table 11.2 Proportion of each product in different room

Room Product	Kitchen	Dining Room	Living Room	Bathroom	Hallway	Bedroom
Smart camera	23%	3%	37%	4%	14%	19%
Temperature/humidity sensor	13%	0%	35%	2%	5%	44%
Smart speaker	35%	3%	23%	11%	1%	27%
Thermostat	8%	7%	17%	2%	19%	47%
Door/window sensor	16%	0%	15%	6%	35%	28%
Motion sensor	21%	1%	7%	28%	31%	12%
Smoke/carbon detector	27%	1%	7%	2%	29%	34%
Smart light	11%	6%	33%	4%	2%	44%
Smart hub	23%	1%	20%	15%	9%	32%
Smart screen	33%	0%	19%	3%	3%	42%
Total	21%	2%	21%	8%	15%	33%

Table 11.3 Rate of each product in different room

Room Product	Kitchen	Dining room	Living room	Bathroom	Hallway	Bedroom
Smart camera	4.17	4.60	4.00	4.57	4.43	4.00
Temperature/humidity sensor	4.27	4	4.36	4.67	4.70	4.61
Smart speaker	4.43	4.45	4.31	4.73	3.67	4.47
Thermostat	3.71	4.14	4.03	4.00	4.20	4.12
Door/window sensor	4.43	–	3.20	4.9	3.89	4.67
Motion sensor	4.13	4	3.82	4.45	4.34	4.13
Smoke/carbon detector	4.37	3.00	3.42	4.5	4.49	3.82
Smart light	4.73	4.38	4.49	4.5	4.86	4.41
Smart hub	3.69	3.00	3.82	3.43	3.50	3.71
Smart screen	4.08	3.67	4.06	4	2.60	4.16
Total	4.20	3.92	3.95	4.37	4.07	4.21

2. To monitor the situation of pets such as cats and dogs.
3. Usually, the smart camera has a wide-angle lens.

When it is placed in the living room, we can see part of the kitchen, dining room, or bedroom at the same time. There are some problems with smart camera in the living room:

1. When stray animals pass by the yard outside the living room, they will accidentally trigger the alarm;
2. Although the monitoring has a wide angle, sometimes people are far away, and the camera cannot successfully perform face recognition.

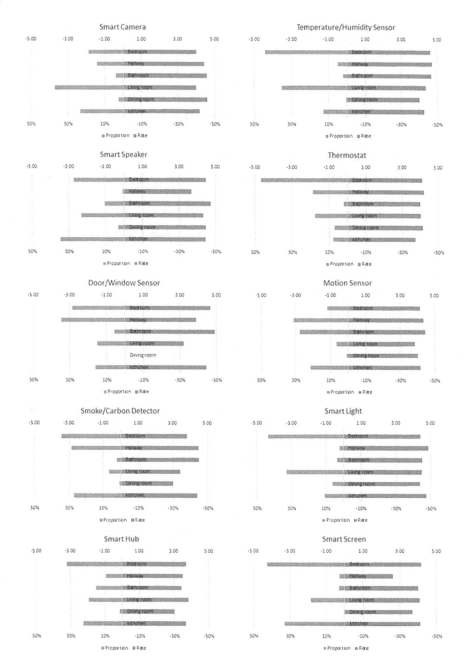

Fig. 11.2 Relation of portion and rate

Thermostat and temperature/humidity sensor are usually used together, and most often place them in the bedroom. The reasons why people most often place them in the bedroom are as follows:

1. People spend a long time staying in the bedroom.
2. People like to control the temperature at bedtime.
3. The temperature of the bedroom is usually different from the temperature of other rooms, and the temperature requirements for sleep are the most rigid.

There are some problems with them in the bedroom:

1. With the help of these products, people still cannot control the temperature to sleep well.
2. Using these products will pay a higher electricity bill.

The reasons why people most often place smart speaker in the kitchen are as follows:

1. People want to have kitchen entertainment while cooking.
2. When cooking, both hands are occupied.

Putting smart speakers in the kitchen allows people to use speech to control home appliances. There are some problems about smart speaker in the kitchen:

1. Smart speakers failed to achieve a seamless connection between room and room.
2. The artificial intelligence of smart speakers is not smart enough to affect the experience.

The reason why people most often place door/window sensor and motion sensor in the hallway is to monitor whether there is an abnormal opening. One problem is that as the door or window is opened and closed many times, the product may fall.

The reasons why people most often place smoke/carbon detector in the bedroom are as follows:

1. If a fire occurs while sleeping, the detector can wake people up in time.
2. Some detectors have a night light feature.

There are some problems with smoke/carbon detector in the bedroom:

1. There was no disaster but the alarm ring, interrupting the user's sleep.
2. If the user smokes, it will also trigger the alarm, which is not friendly to the smoker.

The reasons why people most often place smart light in the bedroom are as follows:

1. People want to avoid the blue spectrum before sleeping.
2. Wake up at night without looking for a light switch in the dark, we can turn the light on and off from the bed.
3. People can make it as a wake-up alarm by setting it up to gradually turn on.

There are some problems with smart light in the bedroom:

1. The noise of the light is not good for relaxation and falling asleep.
2. When multiple smart lights are used at the same time, they often have connection errors.

The reasons why people most often place smart screen in the bedroom are as follows:

1. People use them as alarm clocks, speakers, or making video calls.
2. People use them to voice control other smart products.

There is a problem with smart screen in the bedroom: privacy issues. Some users suggest that it needs a physical camera cover so we can be assured of privacy in the bedroom.

Because most of the smart products are placed in the bedroom, people tend to put the hub in the bedroom.

As shown in Fig. 11.3, the position of AI products in the building space could reflect the needs of modern living space. Many people put door/window sensor and motion sensor in the hallway, which means we should enhance the sense of security when we design it. Many people put smart speaker and smart screen in the kitchen, which suggests us to combine entertainment devices with the kitchen. Another way is to merge the kitchen with the living room so that people can use the entertainment devices in the living room while cooking. Many people put smart camera in the living room, so our design should try to avoid blind spots in the living room. Many people put smart light in the dining room, which may suggest us to make the dining

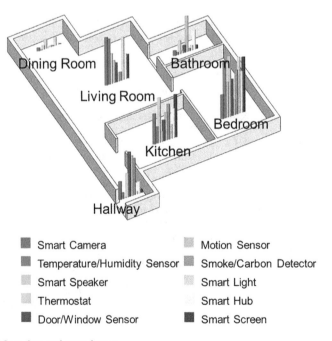

Fig. 11.3 AI product and room layout

room more stylish. Many people use motion sensor to control the lights in the bathroom. It would be helpful if we have more natural lighting in the bathroom. People often place thermostat, temperature/humidity sensor, and smart light in the bedroom, which reminds us that we should consider the energy consumption in the bedroom.

11.4 Conclusion

This research scraped smart building product information by using Gooseeker on Amazon.com to reveal the relationship between AI products function and room layout and provide the proportion and rating of AI products in the building. Smart building products are most often placed in the bedroom. Nevertheless, rooms that are frequently placed may not get the highest rating. We also learned about the reasons and shortcomings of people placing a smart product in a room through reviews. As we read these reviews, we found that most of the low ratings are due to the defects of the product itself. But architects can solve some problems through better design and construction.

References

Crisnapati, P. N., Wardana, I. N. K., & Aryanto, I. K. A. A. (2016, May). Rudas: Energy and sensor devices management system in home automation. In *2016 IEEE region 10 symposium (TENSYMP)* (pp. 184–187). IEEE.

El Mougy, A., Khalaf, A., El Agaty, H., Mazen, M., Saleh, N., & Samir, M. (2017, Oct). Xenia: Secure and interoperable smart home system with user pattern recognition. In *2017 international conference on Internet of Things, embedded systems and communications (IINTEC)* (pp. 47–52). IEEE.

Guo, X., Shen, Z., Zhang, Y., & Wu, T. (2019). Review on the application of artificial intelligence in smart homes. *Smart Cities, 2*(3), 402–420.

Hsu, Y. L., Chou, P. H., Chang, H. C., Lin, S. L., Yang, S. C., Su, H. Y., Chang, C. C., Cheng, Y. S., & Kuo, Y. C. (2017). Design and implementation of a smart home system using multisensor data fusion technology. *Sensors, 17*(7), 1631.

Jakkula, V. R., & Cook, D. J. (2011). Detecting anomalous sensor events in smart home data for enhancing the living experience. *Artificial Intelligence and Smarter Living, 11*(201), 1.

Senarathna, S. S., Muthugala, M. V. J., & Jayasekara, A. B. P. (2018, Sept). Intelligent robot companion capable of controlling environment ambiance of smart houses by observing user's behavior. In *2018 2nd international conference on electrical engineering (EECon)* (pp. 124–129). IEEE.

Shao, P. (2015, Nov). Intelligent control in smart home based on adaptive neuro fuzzy inference system. In *2015 Chinese automation congress (CAC)* (pp. 1154–1158). IEEE.

Wilson, C., Hargreaves, T., & Hauxwell-Baldwin, R. (2015). Smart homes and their users: A systematic analysis and key challenges. *Personal and Ubiquitous Computing, 19*(2), 463–476.

Part IV
Building Performance and Design Evaluation

Chapter 12
A Comparative Study on the Performance of BIPV and Mainstream Insulation Materials for the Energy-Efficient Renovation of Existing Residential Buildings in China

Jing Wang, Xuan Ge, Yijia Miao, and Siu Yu Stephen Lau

12.1 Introduction

With the advent of climate warming, contradiction between energy supply and demand has become increasingly prominent. In light of the Paris Agreement objectives, minimizing energy demand has become a global effort. The renovation of existing buildings plays a fundamental role in meeting sustainable development goals as it has multiple impacts on environment, society, and economy. The key role of building energy-efficiency renovation is that of balancing energy consumption and global warming. The energy consumption at a building scale depends on the design of the building, the usage of the building and the climate zone, etc. (Braulio-Gonzalo & Bovea, 2017). The reasons for building energy losses are either high heat transfer coefficient of the envelope structure, low air tightness of windows, and thermal bridges. Taking the China HSCW-hot summer and cold winter climatic zone as the site, this paper focused on the effectiveness and life cycle performance of the selected envelope construction for energy efficient renovation of residential buildings. Figure 12.1 shows the simplified climatic map of China.

J. Wang · X. Ge
South China University of Technology, Guangzhou, China

Y. Miao (✉)
SOSArchitecture Urban Design Studio, Sheung Wan, Hong Kong, China

University of Hong Kong, Pok Fu Lam, Hong Kong

S. Y. S. Lau
University of Hong Kong, Pok Fu Lam, Hong Kong

Beijing University of Civil Engineering and Architecture, Beijing, China

Climate Zones	Mean Monthly Temperature	
	Coldest	Hottest
Severe Cold	≤ -10°C	≤ -25°C
Cold	-10-0°C	18-28°C
HSCW	0-10°C	25-30°C
HSWW	> 10°C	25-29°C

Fig. 12.1 Chinese climate zones. (Sources: Authors)

Exterior insulation and BIPV are two of the most obvious solutions for innovating energy efficiency of residential buildings. Several studies on incorporating insulation materials on building envelope showed great potential for reducing energy consumption. Zhang et al. (2019) analyzed the influence of external insulation by the material property and thickness based on energy consumption simulation and life cycle economic evaluation. The result shown that adding insulation to exterior walls has less effect on the cumulative annual cooling load, but a greater impact on heat load for those in the cold climate zones. X. Lu and Memari (2019) studied building energy performance with different types of insulation materials. Their simulation results shown that adding insulation materials could significantly improve building energy performance. Polyisocyanurate performs best among conventional insulating materials, with an annual energy savings of 37% in the case of Pittsburgh. Kumar et al. (2020) investigated the optimal insulation thickness of different building materials through the degree-days method and life-cycle economic analysis and reported the payback periods and carbon mitigation of earthen and mud concrete materials.

Compared with enhanced insulation, BIPV also emerges to be an ideal solution since it converts building envelopes into localized powerhouses. Thus, energy generated can supply the building operation while replacing building components. Saretta et al. (2019) conducted a systematic analysis of the literature on energy-saving renovation of buildings in various urban contexts, and substantiated the significance of BIPV system for urban energy transition. Aguacil et al. (2019) presented a methodology to select appropriate BIPV components in renovation projects based on autonomous-consumption and autonomous-sufficiency. Evola and Margani (2016) investigated the energy and economic performance of renovating residential buildings in the temperate climate zone by integrating photovoltaic panels on façades. The result has shown that lower BIPV component price, higher photovoltaic efficiency, and higher autonomous-consumption rate could enhance economic profitability. Saretta et al. (2019) reviewed the literature of BIPV in the renovation of building envelope, and argued that the application of BIPV could fulfill several other tasks as a building façade such as glare protection and solar protection when designed properly. In summary, the application of BIPV is relevant for building envelope, economic advantage, and environmental impact.

With the proliferation of green building evaluation standards, building energy efficiency codes, thermal regulatory requirements of building envelopes have gradually increased (Lu et al., 2020). However, as the statutory criteria are more concerned about the improvement of thermal performance, it has become necessary to study and explore envelope optimization method that has better performance on energy-saving, economic, and environmental benefits. At present, most scholars have used simulation software to select and optimize building envelope by incorporating air layer (Zhang & Yang, 2019), phase change materials (PCM) (Jiang et al., 2017; Mannivannan & Ali, 2015), as well as the use of photovoltaic curtain wall (Gonçalves et al., 2021; Huang et al., 2018), etc. However, there is a lack of data on the energy performance, economic analysis, and sustainability in comparing enhanced insulation and BIPV. This paper has modeled and analyzed two envelope optimization approaches from the perspective of energy performance, economic performance, life cycle performance, and overall potentials. The results provide a reference for the design and evaluation of green building envelope and benefit stakeholders in making better choices based on specific requirements. Figure 12.2 shows the design process of this research.

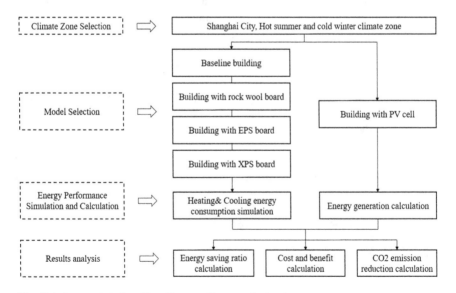

Fig. 12.2 Research design: Flow diagram. (Sources: Authors)

12.2 Methodology

12.2.1 Model Selection

The exterior wall is an important objective in building energy-saving renovation. Currently, thermal performance is mainly achieved by adding external thermal insulation materials to reduce the building energy load. More recently, the maturity of BIPV technology and market has offered building exterior wall an added-ability to produce electricity, to make up for various defects from the use of fossil energy. In order to clarify the energy efficiency and economics of these two renovation strategies, enhanced thermal insulation and BIPV for building facades, this paper has selected the hot summer and cold winter climate zone for the investigation. Shanghai City has been chosen for it presents the highest population density in a climatic zone which according to the national planning does not benefited from district heating provided and managed by the government for the public. As a result, inhabitants in Shanghai and this climate zone end up paying for their own heating means in the winter months – largely by heating provided by household air conditioning units. For the HSCW climate zone, thermal performance of external walls is particularly important for building energy consumption. Table 12.1 shows the key design indicators and parameters regulated by "Design Standard for Energy Efficiency of Residential Buildings in Hot Summer and Cold Winter Zone JGJ 134-2010." Based on the local standard, the authors have selected compatible PV modules and available external insulation construction for the quantitative study on their applications on existing residential buildings in the climate zone.

12.2.1.1 Selection of PV and Insulation Materials

Regarding the choice of photovoltaic modules, the current market is ambulant with various commercialized BIPV products with varying properties and prices. With reference to the 2019 version of China PV Industry Development Roadmap,

Table 12.1 Key design indicators and parameters for residential buildings in hot summer and cold winter zone

Indicator	Shape coefficient	Exterior wall heat transfer coefficient($W/m^2 \cdot K$)	Window-wall ratio of south wall
Parameter	0.55(3 floors) 0.40(4–11 floors) 0.35(more than 12 floors)	1.5(shape coefficient ≤ 0.4, thermal inertia > 2.5) 1.0(shape coefficient ≤ 0.4, thermal inertia ≤ 2.5) 1.0(shape coefficient > 0.4, thermal inertia > 2.5) 0.8(shape coefficient > 0.4, thermal inertia ≤ 2.5)	0.45

Data source: Design Standard for Energy Efficiency of Residential Buildings in Hot Summer and Cold Winter Zone JGJ 134-2010

monocrystalline and polycrystalline silicon cells are considered the main types of photovoltaic cells. The low photoelectric conversion efficiency and high production costs of thin-film cells still restrict their large-scale application. Table 12.2 shows key information of monocrystalline and polycrystalline silicon cells, from which one could see that monocrystalline silicon cells produce more electricity at the same cost and effective life as polycrystalline silicon cells. Accordingly, this paper chooses monocrystalline silicon cells as the PV material to be studied.

For the selection of exterior insulation materials, refer to the most authoritative and relevant domestic "Design Standard for Energy Efficiency of Residential Buildings in Hot Summer and Cold Winter Zone JGJ 134-2010, Technical Regulations for Energy-saving Reconstruction of Existing Residential Buildings JGJ/T 129-2012, Technical Standards for External Insulation Engineering of External Walls JGJ 144-2019" and other industry standards; the current mainstream external wall insulation materials are mainly expanded polystyrene panel (EPS), extruded polystyrene panel (XPS), mineral binder and expanded polystyrene granule bonding plaster, thermal insulation slurry, rigid polyurethane foam board (PUR /PIR), and other organic insulation materials. Taking into account the increasing importance of fireproof performance of external wall insulation materials, inorganic thermal insulation materials have also begun to occupy a certain market share. The most representative ones are rock wool board, expanded perlite, and foamed concrete, etc.

With reference to relevant standards, this paper has selected 120 mm thick rock wool board, 100 mm EPS board, and 100 mm XPS board as the external wall insulation materials for the analysis and comparison. The key parameters of these three types of external wall insulation materials are shown in Table 12.3.

Table 12.2 Key information of the two kinds of PV cells

PV cell	Average incident photon-to-electron conversion efficiency(%)	Module power (W)	Normal module cost(CNY/W)	Effective life (Year)
Monocrystalline silicon	22.3	320–360	1.31	20
Polycrystalline silicon	19.3	285–330	1.31	20

Data sources: 2019 version of China PV Industry Development Roadmap by China Photovoltaic Industry Association and CCID thiktank、Supplementary notice to several Suggestions on Promoting the Healthy Development of Non-water Renewable Energy Power Generation by Ministry of Finance of the People's Republic of China. The Notice on exerting the role of price leverage to promote the healthy development of the photovoltaic industry issued by National Development and Reform Commission

Table 12.3 Key information of the three exterior wall insulation materials

Insulation material	Thickness(mm)	Thermal conductivity(W/ m·K)	Heat transfer coefficient(W/ m²·K)	Production cost(CNY/ m³)	Effective life(year)
Rock wool board	120	0.038	0.32	500	25
EPS board	100	0.039	0.39	400	25
XPS board	100	0.029	0.29	800	25

Data sources: Document research, Market survey, Technical Standards for External Insulation Engineering of External Walls JGJ 144-2019

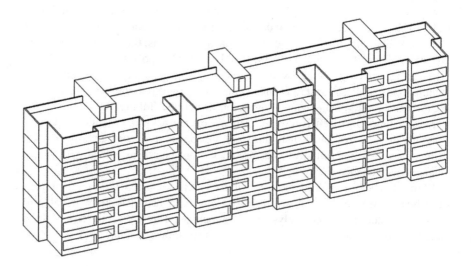

Fig. 12.3 3D model of the case building. (Sources: Authors)

12.2.1.2 Selection of Case Building

Based on the relevant provisions of "Design Standard for Energy Efficiency of Residential Buildings in Hot Summer and Cold Winter Zone JGJ 134-2010," a case building for simulation is built, as shown in Fig. 12.3. The case building is located in Shanghai, typically 6 floors high with the standard floor height of 3.3 m. The total building area is 4806 square meters, the shape coefficient is 0.33, and the window-wall ratio of south wall is 0.35. The envelope information is shown in Table 12.4. In the subsequent simulation and comparative analysis, relevant parameters of the roof, exterior windows, and ground floor will not be altered, only the exterior wall (envelope) will change accordingly with the changes of added materials. Also, when calculating the power generation of PV system, it is assumed that the building is free from solar exposure since there are no neighbors in the vicinity.

Table 12.4 Key information of the case building

Building envelope	Heat transfer coefficient(W/m²·K)	Construction
Exterior wall	1.70	8 mm exterior wall tiles surface + 5 mm cement mortar + 200 mmconcrete + 15 mm cement mortar
Exterior window glass	1.96	12 mm single layer tempered glass
Exterior window Proximate matter	3.63	Aluminum alloy
Roof	3.64	20 mm cement mortar + 120 mm concrete +20 mm cement mortar excluding waterproofing and thermal insulation
First floor	3.10	100 mm concrete +20 mm cement mortar + 40 mm granite

12.2.2   Energy Performance Simulation and Calculation

Professional simulation software is used to study the energy performance stipulated by design standards and codes applied on the case building with the three different kinds of external insulation materials, and the case building with photovoltaic modules. It should be emphasized that (1) the application of thermal insulation materials on the outer wall will mainly affect the energy requirement for indoor heating and cooling. As explained, the research looks into the impacts on energy consumption during the winter months; (2) For the purpose of measuring the performance at its best orientation, the south faced outer wall is used in the simulation for comparison of the BIPV versus the corresponding faced with thermal insulation (Figs. 12.4 and 12.5).

Since photovoltaic products have little effect on the thermal insulation performance on the building envelope, the following energy performance simulation will not consider the thermal insulation performance due to the photovoltaic modules.

In this paper, Energy plus 9.1.0 version, which was jointly developed by the US Department of Energy (DOE) and the Lawrence Berkeley State Key Laboratory, is used for the building energy consumption simulation. The simulation makes reference to the "Design Standard for Energy Efficiency of Residential Buildings in Hot Summer and Cold Winter Zone JGJ 134-2010," which regulates both heating and cooling needs in this area. The internal design temperature of the building is 18 degrees Celsius in winter and 26 degrees Celsius in summer; observing that the heating calculation period is from 1st December of the current year to 28th February of the following year. The cooling calculation period is from 15th June to 31st August of the current year; outdoor meteorological parameters adopt the typical weather year; for heating and air conditioning, the number of air changes is 1.0

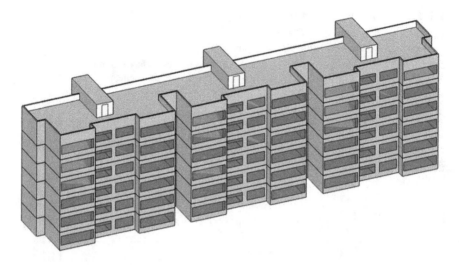

Fig. 12.4 3D model of the case building with exterior insulation material (south façade). (Sources: Authors)

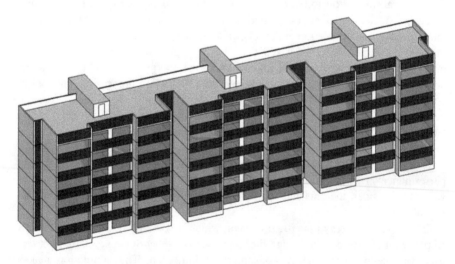

Fig. 12.5 3D model of the case building with PV modules (south façade). (Sources: Authors)

times/hour; the average indoor heat gain being 4.3 W/m². Other parameters related to indoor energy consumption followed default setting.

In addition, this paper uses PVsyst software, which is generally recognized by the academic circle to calculate the photovoltaic power generation of the model. The photovoltaic module uses 60-pieces with a size of 0.996 m × 1.66 m each. As a result, the total number of photovoltaic panels that could be installed on the case building is 456, with a total area of 726.68 square meters. The formula for calculation of photovoltaic power generation is as follows:

$$\text{Power generation due to photovoltaic system} =$$
$$\left(\begin{array}{l} \text{radiation intensity of receiving surface from April to September} \times \\ \text{area of photovoltaic panel} + \text{radiation intensity of receiving surface} \\ \text{from October to March} \times \text{area of photovoltaic panel} \end{array} \right) \times (12.1)$$
$$\text{discharge efficiency of battery} \times \left(1 - \text{loss rate of photovoltaic system} \right)$$

12.2.3   Cost-Benefit Analysis

Based on the cost per unit area of PV and insulation materials and the external skin information of the case building given above, the initial investment required when using various materials have been calculated. Based on the national standard for design of civil buildings GB 50352-2019, the service life of the building is assumed as 50 years, the replacement cost of various materials is thus being taken into account. Regarding the maintenance cost of these two approaches, while the PV industry has adopted the concept of system life cycle investment cost and related algorithms represented by LCOE, the operation and maintenance cost due to external wall insulation materials has not yet been estimated properly or systematically. For this reason, the research has not considered the cost impact for maintenance.

In addition, the research has adopted the cost benefit study to account for indirect economic benefit if any due to BIPV whose main feature is to generate electricity instead of proving thermal buffer that result in energy saving, as in the case of the benefit of using thermal insulation material. Here, the cost-benefit ratio refers to the initial investment required when a certain material is used to produce an energy saving rate of 1%. The calculation method is:

$$\text{Cost} - \text{benefit ratio} = \text{Initial investment} / \left(\text{energy saving ratio} \times 100\% \right) \quad (12.2)$$

12.3   Results and Discussion

12.3.1   All Year Energy Performance

This paper uses energy saving ratio to represent the energy performance of the simulated buildings with different materials, which can be calculated by the following formula:

$$\text{Energy saving ratio} = \left(\begin{array}{l} \text{Energy consumption of the baseline building } - \\ \text{energy consumption of the building with one specific material} \end{array} \right) \quad (12.3)$$

/energy consumption of the baseline building.

The heating, cooling and total energy consumption data of the heating, ventilation and air-conditioning (HVAC) system for the baseline building and the case building with 120 mm rock wool board, 100 mm EPS board, and 100 mm XPS board on the south outer wall are shown in Fig. 12.6.

The heating, cooling, and total energy saving ratio of HVAC system for the buildings with 120 mm rock wool board, 100 mm EPS board, and 100 mm XPS board on the southern outer wall are shown in Fig. 12.7.

The total energy saving ratios contributed by different materials based on the whole year's overall data are shown in Table 12.5.

The data shows that when thermal insulation materials are installed only on the south facing outer wall, its energy-saving effect on the HVAC system is very limited. Its energy-saving rate for the HVAC system is between 2.26% and 2.40%. Specific to the different functions of the HVAC system, installing thermal insulation on the south facing wall reduces the energy consumption of building cooling by a large margin, and the energy saving rate exceeds 5%, up to 5.36%. Simultaneously, its impact on building heating energy consumption is minimal, with the highest energy saving rate being 1.22% and the lowest being 1.13%. Among these three kinds of external insulation materials, the rock wool board and XPS board have the same energy-saving effect on the building heating system, and the EPS board has the worst energy-saving effect.

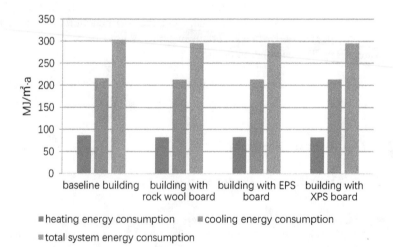

Fig. 12.6 Heating, cooling, and total data of the HVAC system energy consumption from the four simulated models. (Sources: Authors)

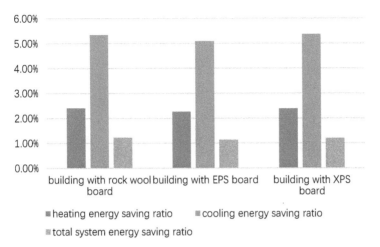

Fig. 12.7 Heating, cooling, and total energy saving ratio of the HVAC system from the three simulated models with different insulation materials. (Sources: Authors)

Table 12.5 Annual data of the HVAC system energy consumption and energy saving ratio from the four simulated models

Building model	Energy consumption of the HVAC system(kWh/m²·a)	Energy saving ratio of the HVAC system (%)
Case building	83.98	0
Building with rock wool board	81.96	2.40
Building with EPS board	82.08	2.26
Building with XPS board	81.98	2.39

Based on the data analysis of PVsyst (student version), when the installation inclination of photovoltaic panels in Shanghai is 90 degrees, from October to March, the ratio of solar radiation loss relative to the horizontal plane is 22.2%, and the effective radiation intensity is 468 kWh/m²; From April to September, the solar radiation loss ratio relative to the horizontal plane is 54.9%, and the effective radiation intensity is 362 Wh/m² (as shown in Fig. 12.8).

The battery discharge efficiency is 80%, and the photovoltaic system loss rate is 20%. As a result, after installing the photovoltaic system, the building's power generation from October to March was 48537.11 kWh, from April to September was 37543.66 kWh, and the total annual power generation was 86080.77 kWh. The power of the PV system can be calculated with the followed formula:

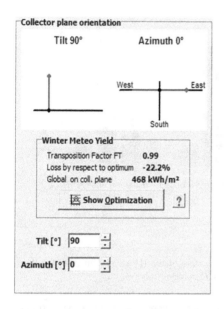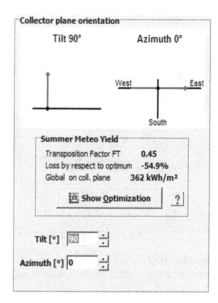

Fig. 12.8 Solar radiation intensity received by PV in the building model. (Sources: Authors, computed by PVsyst student version)

$$\text{Power of the } PV \text{ system} = \text{Annual power generation } /$$
$$\left[\begin{array}{c} \text{Annual effective hours} \times \text{Battery discharge efficiency} \times \\ \left(1 - PV \text{ system loss rate}\right) \end{array} \right] \qquad (12.4)$$

Based on the total building area, the equivalent power generation is 17.91 kWh/$(m^2 \cdot a)$[1], and the energy saving rate for the HVAC system is 21.3%.

12.3.2 Economic Performance

In the case building, the surface area of the south outer wall where insulation materials need to be added is 974.88 m^2. When 120 mm thick rock wool board is used, its total volume is 116.99 m^3 and the cost is 58,500 CNY; when 100 mm thick EPS board is used, its total volume is 97.49 m^3 and the cost is 39,000 CNY; when a 100 mm thick XPS board is used, its total volume is 97.49 m^3, and the cost is 78,000 CNY. Based on the above stated life of the insulation material, each insulation material needs to be replaced once. As a result, the total cost of the rock wool board equates 117,000 CNY, the total cost of the EPS board equates 78,000 CNY, and the total cost of the XPS board is 156,000 CNY. Meanwhile, the total power generated

---

[1] equivalent power generation = total annual power generation / total building area

per module of the photovoltaic system is 122.2 kW, and the module cost of the entire photovoltaic system is 160178.7 CNY. The photovoltaic system needs to be replaced twice during the designed life of the building, so its total cost becomes 480,536.1 CNY. As a renewable energy source, photovoltaic power generation is qualified to receive financial support from the national and local governments. According to the "Shanghai Special funds for Renewable Energy and New Energy Development (version 2020)," the subsidy unit rate for distributed photovoltaics (including household photovoltaics) equates 0.15 CNY/kWh (for the period of 2020 to 2028). Based on the above simulation results, the total power generation of the photovoltaic panels installed is 86,080.77 kWh per year, so that the governmental subsidy for photovoltaic panels will be 103,296.92 CNY for the period of 2020–2028. Taking the government subsidies into account, the total cost of installing photovoltaics shall be 377,239.18 CNY.

According to the formula (12.2), the cost-benefit ratios of the four different materials are shown in Table 12.6. No matter what kind of external wall insulation material is compared, the net contribution of PV (measured by cost-benefit ratio) is at a pronounced advantage.

12.3.3   Sustainability Analysis

In addition to energy and economic benefits, carbon reduction benefits have emerged as the one that draws global attention in terms of climate change reactions emission factors are used to check emissions. . According to relevant studies, residents in Shanghai area mainly use distributed air conditioners to heat and cool their dwelling spaces, which are powered by electricity (Guo et al., 2014; Song et al., 2017; Zhou et al., 2013). To calculate the amount of CO_2 emission reduction from the buildings with different materials with a simplified method, the following formulas are used:

$$\text{Amount of } CO_2 \text{ emission reduction from electricity} =$$
$$\text{Amount of electricity reduction} \times \text{electricity emission factor} \qquad (12.5)$$

With reference to the "Shanghai Greenhouse Gas Emission Accounting and Reporting Technical Document SH/MRV-001-2012" issued by the Shanghai Municipal Development and Reform Commission, the default value of electricity emission factor is 7.88tCO2/104 kWh. Therefore, when using photovoltaic systems, the carbon reduction amount can be calculated by multiplying the photovoltaic system power generation and the default value of the electricity emission factor. As

Table 12.6 Cost-benefit ratio of the four materials

Material	Rock wool	EPS	XPS	BIPV
Cost-benefit ratio (CNY)	48750.00	34513.27	65271.97	17,710.76

Data source: Authors

a result, the case building could reduce CO2 emissions by up to 67.83t annually and reduces CO2 emissions by 3391.58t during the entire life cycle.

On the other hand, when the rock wool board is used, the building can save 9706.23 kWh of power every year, reducing CO2 emissions by 7.65t per year and reducing CO2 emissions by 382.43t in the whole life cycle. When the EPS board is adopted, the building can save 9132.13 kWh of power per year, thereby reducing CO2 emissions by 7.20t per year and reducing CO2 emissions by 359.81t during the entire life cycle. When the XPS board is used, the building can save 9652.82 kWh of power each year, thereby reducing CO2 emissions by 7.61t and reducing CO2 emissions by 380.32t during the entire life cycle. The CO2 emission reductions of the four options during the entire building life cycle are shown in Fig. 12.9.

It can be seen that the PV system has an obvious advantage.

When studying the relationship between the thickness of various external wall insulation materials and the overall benefits, many scholars have pointed out that after the heat transfer coefficient of insulation materials reaches a certain limit, its impact on the energy-saving effect of buildings hardly changes (Kumar et al., 2020; Wei et al., 2014; L. Zhang et al., 2019). Blindly increasing the thickness of insulation materials will make the growth rate of input costs far exceed that of the growth rate of energy-saving benefits, which becomes very uneconomical, so that the optimal value of its cost-benefit ratio is basically fixed.

Today BIPV is considered as one of the most important technologies, by which the standards for net zero energy buildings are becoming achievable. BIPV can be used in new buildings as well as in existing buildings that are being renovated, resulted in advantages as demonstrated by the case building. According to trend forecasts, by 2023, the cost-benefit ratio of photovoltaic modules can be reduced by about 19.8%. With the implementation of government economic incentive policies, the continuous improvement of photovoltaic conversion efficiency, the increase of

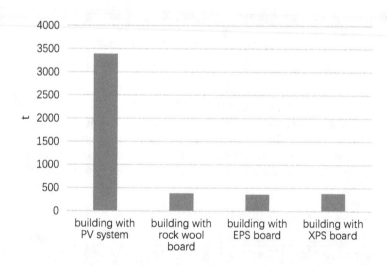

Fig. 12.9 Lifecycle amount of CO_2 emission reduction of the four materials. (Sources: Authors)

effective service life and the reduction of module production costs, the cost-benefit ratio of photovoltaic power generation will gradually decrease, and the competitiveness of photovoltaic power generation will be greatly improved in the foreseeable years (Fig. 12.10).

Moreover, the application of photovoltaic components is an important solution for the construction industry to achieve low-carbon, carbon peaking, and even carbon-neutral national goals. It can fully enjoy dividends brought by green finance and become an important market economy support mechanism, enhancing the comprehensive benefits of photovoltaic systems. More importantly, the application of photovoltaic components can fundamentally improve society's energy supply mechanism, making the whole energy production, transportation and application chain cleaner, environmentally friendly, and sustainable, and providing a reliable guarantee for the healthy operation of the society.

12.4 Conclusion

In this paper, a detailed assessment of the optimal performance of the enhanced insulation and BIPV by means of simulations was presented. Four types of exterior wall materials (rock wool, EPS, XPS, BIPV) were assessed. The assessment shows that the photovoltaic system cost-benefit ratio is lower than that of insulation materials, while photovoltaic system has an obvious advantage as discussed. Simultaneously, the comprehensive economic performance of the photovoltaic system as a building envelope material shows a continuous improvement trend. For the early realization of near-zero energy consumption, carbon peaking, carbon neutrality, and other field goals, increasing the application scale of BIPV components in the construction industry seems also a meaningful path. This research could not only help users and architects to acknowledge BIPV system as a suitable option for

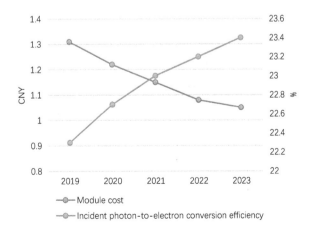

Fig. 12.10 Trends forecast of the module cost and incident photon-to-electron conversion efficiency from 2019 to 2023. (Sources: Authors, based on data from the 2019 China Photovoltaic Industry Development Roadmap)

building skins in the HSCW Climate Zone in China but also help governments or decision-makers to promote appropriate technologies by policies and incentives.

Acknowledgments This paper is funded by the Open Project (International Cooperation) of the State Key Laboratory of Subtropical Building Science, South China University of Technology (Project name: Research on the application of photovoltaic system in curtain wall renovation, Project number: 2019ZA03)

Appendix

Formulas

- **Power generation due to photovoltaic system** = (radiation intensity of receiving surface from April to September× area of photovoltaic panel + radiation intensity of receiving surface from October to March × area of photovoltaic panel) × discharge efficiency of battery × (1 − loss rate of photovoltaic system)
- **Cost-benefit ratio** = Initial investment / (energy saving ratio × 100%)
- **Energy-saving ratio** = (Energy consumption of the baseline building − energy consumption of the building with one specific material) / energy consumption of the baseline building.
- **Power of the PV system** = Annual power generation/ [Annual effective hours × Battery discharge efficiency × (1 − PV system loss rate)]
- **Equivalent power generation** = total annual power generation / total building area
- **Amount of CO_2 emission reduction from electricity** = Amount of electricity reduction × electricity emission factor

References

Aguacil, S., Lufkin, S., & Rey, E. (2019). Active surfaces selection method for building-integrated photovoltaics (BIPV) in renovation projects based on self-consumption and self-sufficiency. *Energy and Buildings, 193*, 15–28.

Braulio-Gonzalo, M., & Bovea, M. D. (2017). Environmental and cost performance of building's envelope insulation materials to reduce energy demand: Thickness optimisation. *Energy and Buildings, 150*, 527–545.

Evola, G., & Margani, G. (2016). Renovation of apartment blocks with BIPV: Energy and economic evaluation in temperate climate. *Energy and Buildings, 130*, 794–810.

Gonçalves, J. E., van Hooff, T., & Saelens, D. (2021). Simulating building integrated photovoltaic facades: Comparison to experimental data and evaluation of modelling complexity. *Applied Energy, 281*, 116032.

Guo, C., Yan, D., Peng, C., Cui, Y., & Zhu, A. (2014). Research on the current situation of Shanghai residential heating in winter heating. *Ventilation and Air Conditioning, 6*, 11–15.

Huang, J., Chen, X., Yang, H., & Zhang, W. (2018). Numerical investigation of a novel vacuum photovoltaic curtain wall and integrated optimization of photovoltaic envelope systems. *Applied Energy, 229*, 1048–1060.

Jiang, G., Huang, J., Liu, M., & Cao, M. (2017). Experiment and simulation of thermal management for a tube-shell Li-ion battery pack with composite phase change material. *Applied Thermal Engineering, 120*, 1–9.

Kumar, D., Zou, P. X., Memon, R. A., Alam, M. M., Sanjayan, J. G., & Kumar, S. (2020). Life-cycle cost analysis of building wall and insulation materials. *Journal of Building Physics, 43*(5), 428–455.

Lei, S., Xiang, Z., Jingsi, Z., Shun, Z., & Shuai, Y. (2017). Heating behavior and energy consumption simulation of residential heat pump type air conditioners in Shanghai. *Heating Ventilating & Air Conditioning, 09*, 55–60.

Lu, X., & Memari, A. (2019). Comparative analysis of energy performance for residential wall systems with conventional and innovative insulation materials: A case study. *Open Journal of Civil Engineering, 9*(3), 240–254.

Lu, S., Wang, Z., & Zhang, T. (2020). Quantitative analysis and multi-index evaluation of the green building envelope performance in the cold area of China. *Sustainability, 12*(1), 437.

Mannivannan, A., & Ali, M. J. (2015). Simulation and experimental study of thermal performance of a building roof with a phase change material (PCM). *Sadhana, 40*(8), 2381–2388.

Saretta, E., Caputo, P., & Frontini, F. (2019). A review study about energy renovation of building facades with BIPV in urban environment. *Sustainable Cities and Society, 44*, 343–355.

Song, L., Zhou, X., Zhang, J., Zheng, S., & Yan, S. (2017). Research on heating behavior and energy consumption simulation of residential heat pump air conditioners in Shanghai Heating. *Ventilation and Air Conditioning, 9*, 55–60.

Standard, C. I. (2010). Design standard for energy efficiency of residential buildings in hot summer and cold winter zone. In *JGJ*.

Wei, H., Liu, J., & Yang, B. (2014). Cost-benefit comparison between Domestic Solar Water Heater (DSHW) and Building Integrated Photovoltaic (BIPV) systems for households in urban China. *Applied Energy, 126*, 47–55.

Zhang, T., & Yang, H. (2019). Heat transfer pattern judgment and thermal performance enhancement of insulation air layers in building envelopes. *Applied Energy, 250*, 834–845.

Zhang, L., Hou, C., Hou, J., Wei, D., & Hou, Y. (2019). Optimization analysis of thermal insulation layer attributes of building envelope exterior wall based on DeST and life cycle economic evaluation. *Case Studies in Thermal Engineering, 14*, 100410.

Zhou, X., Zhang, Q., Zhang, J., Zhang, X., Luo, M., & Zhu, Y. (2013). Winter indoor environment investigation and thermal demand analysis of residential buildings in Shanghai heating. *Ventilation and Air Conditioning, 43*(6), 64–66.

Chapter 13
Performance-Oriented Research on the Design Strategy of Prefabricated Building Facades

Junjie Li, Shuo Tian, and Yuanhui Liu

13.1 Introduction

In terms of building system, the prefabricated building system can be divided into the assembled structure system, the enclosure system, and the assembled interior decoration system (Standard for assessment of prefabricated building, 2017). The assembled enclosure system is the main point to the performance of the prefabricated building (Lin, 2003), which directly affects the heat change, natural ventilation, and the level of natural light of the building from the external environment (Song, 2000). These three aspects will constitute more than 60% of the air conditioning load (Crosbie, 1997). The components of high-performance building enclosing system include energy-saving walls, roofs, floors, and other impermeable forms. High-performance doors, windows, and curtain wall glass can better meet the insulation, insulation, shade, light transmission, ventilation, field of view, and other needs. At the same time, its physical properties can be further improved according to changes in climatic conditions, so as to achieve the goal of maintaining a comfortable thermal environment in the indoor space while reducing the energy consumption of air conditioners (Lin et al., 2015). Therefore, the optimized design of saving building energy and improving building environment performance is the basis and key for the prefabricated enclosure system to achieve high performance and low energy consumption. However, there are great differences in the durability and performance of prefabricated building enclosure system. In terms of the systematic research on the prefabricated building enclosure system, it is an effective way to improve its durability and performance according to the level of industrialization and informatization, the consumption of energy and resource, and the performance of the use of environment.

J. Li (✉) · S. Tian · Y. Liu
School of Architecture and Design, Beijing Jiaotong University, Beijing, China
e-mail: lijunjie@bjtu.edu.cn

© The Author(s), under exclusive license to Springer Nature Switzerland AG 2021
S. S. Y. Lau et al. (eds.), *Design and Technological Applications in Sustainable Architecture*, Strategies for Sustainability, https://doi.org/10.1007/978-3-030-80034-5_13

The characteristic of integration components of the prefabricated building demands the design method with the unit module as the core, and the flexibility, fast installation, and factory prefabricated of the facade unit as the prefabricated building enclosure system are important opportunities for the sustainable development of the prefabricated building (Guo, 2013). The facade is the interface between the building and the external space which can effectively protect against the impact of external climate change on the physical environment of the building interior space. Building facade unit is not only a single building skin interface but also includes a protective structure unit composed of skin, space, and system. As a unique form of enclosure components, the prefabricated building facade unit divides the building skin interface into several module units with a certain module size which are prefabricated in the factory and quickly assembled with the main structure in the field.

13.2 Overview of Existing Research

The research of prefabricated buildings has a long history, and its real development began with the explosive growth of housing shortage after World War II, and the market urgently needed a large number of fast-moving, low-cost housing to fill the demand, which created a lot of opportunities for architects to practice and explore (Knaack et al., 2012). To this day, research on different angles of prefabricated buildings continues. Firstly, in terms of energy and resource utilization based on ecological concepts, the key to the idea of "less cost and more use" advocated by American architect Buckminster Fuller (1936) is to require the integration of each resource input and make it most efficient to use, arguing that "the overall performance is greater than the total." He applied this concept to the factory prefabricated light residential system named Dymaxion which sought to achieve solid and functional building space with minimal material input, and the same material can be built again at the same place to response the goal of recycling resource. At the same time, Dymaxion can effectively reduce building energy consumption through natural lighting and ventilation and under certain conditions can achieve the self-sustainability in energy level through solar energy collection and technology storage (Ni, 2010). Secondly, in terms of open systems based on standardized modules, German architect Konard Wachsmann and American architect Walter Gropius (1945) developed an assembled General Panel System together that breaks buildings into such as internal and external walls, floors, ceilings, roofs, and other composite performance plates with structure and thermal insulation carefully designed the node systems between plates to meet the needs of rapid construction and system expansion (Chen, 2011; Zhu, 2013). Wachsmann also uses space steel mesh frame systems for integrated installation of building and structural and equipment systems in large-span prefabricated buildings (Lin et al., 2017). Influenced by Wachsmann, German architect Fritz Haller (1976) developed an assembled USM steel modular system based on the versatility and scalability of the full life cycle of building, included MAXI, MIDI, MINI, and modular furniture. The first three corresponds to the plant, multi-story residential, and small residential (Haller & Wichmann, 1989; Hovestadt & Hovestadt, 1999). Using the history experiences of manufacturing, American architects Stephen Kieran and James Timberlake (2009) have been working to introduce the concept of Mass Customization into the design and construction of assembled aluminum structures in recent years (Kieran & Timberlake, 2003). Thirdly, after the product system, a series of studies have been made on the physical performance of comfort and energy-saving requirements. Fabian Ochs et al. have developed prefabricated wood-enclosed enclosure modules that integrate active equipment such as micro-heat pumps and all-heat switches to meet the building's energy efficiency and comfort from a performance point of view while saving construction time and manpower (Zimmermann, 2009; Konstantinou & Knaack, 2013).

Thus, the point in the research of prefabricated building is not only focus on how to prefabricate and assemble the building system but also on the integrated design opportunities and platforms it provides to realize a deeper pursuit of building sustainability: more efficient utilization of energy and resource, more open building systems, and more comfortable and energy-saving physical performance.

13.3 Research Objects and Methods

According to the three factors of material specification, transportation restriction and ergonomics, this study takes the sustainable performance as the starting point, takes the construction strategy and energy consumption simulation comparison of the facade unit of prefabricated building as the goal, and selects the facade unit with height of 3000 mm and width of 3000 mm as the basic function module and the basic module whose length is determined according to the actual demand as the research object. For the reason that the common building size takes 3M as the modulus, most of them are 3M, 3.3 m, 3.6 m, 3.9 m, etc., so this study adopts the module of 3M as the research object. The facade module of the study meets the demand of the standardized modules in office, medical, residential, and other buildings, so that the research results match the actual application scenario (Fig. 13.1). According to the depth of the facade unit and the construction characteristics, this study discusses the construction and connection of the prefabricated facade unit with sustainability included cavity type, well box type, and embedded type as objects as the guide and verifies the application effect of the practice in different climate zones through software simulation.

13.3.1 The Design Strategy of the Three Facade Units

Cavity design strategy is a double-layered surface form structure on a larger scale; the spatial scale can be between 600 mm-1500 mm, forming a space similar to the sunspace which allows people to move in the room (Fig. 13.2). The well box design strategy is the construction of a double-layered surface form, using the air layer

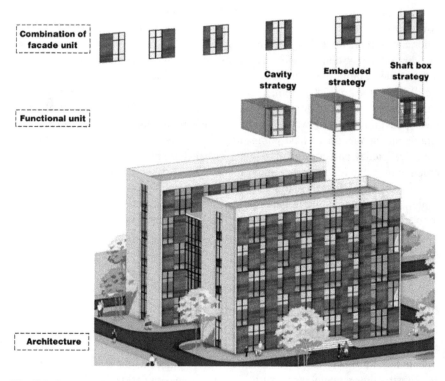

Fig. 13.1 Facade unit scenario analysis

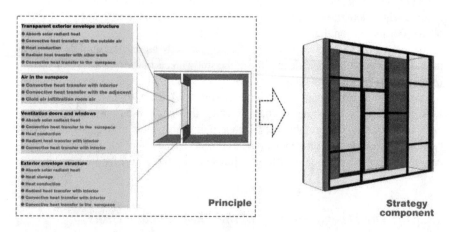

Fig. 13.2 The design of cavity strategy construction process

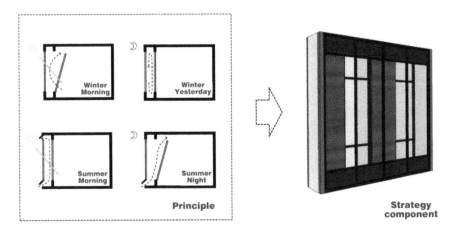

Fig. 13.3 The design of shaft box strategy construction process

between the skins for direct or indirect energy exchange and transmission with the interior, with a spatial scale between 150 mm and 450 mm (Fig. 13.3). The embedded design strategy is that the facility of energy collection and radiation is embedded inside the wall or attached to the wall surface, and the internal heat source is divided into gas and liquid (Fig. 13.4a, b).

13.3.2   The Construction and Connection of the Three Facade Units

13.3.2.1   Cavity Strategy Type

In order to avoid the double-pillar problem between the same layer in cavity design strategy module when it is put together, the module design uses the C-type steel column to connect module so the adjacent modules can buckle and be bolted by a pre-welded nut on the outside of the C-type steel. In order not to affect the C-type steel column buckle connection, the above I-steel beam by the misplacement way to reserve space for adjacent modules. The I-steel of the outer connecting part of the upper and lower facade units is covered with aluminum plate and rock wool insulation to solve the thermal bridge problem, as shown in Fig. 13.5. In order to prevent the hot and cold bridge effect of the C-type steel, add insulate rubber sheets between the C-type steels on the buckle and fill XPS inside to block heat transfer between the steels. When the side plasterboard of module partition is installing, it needs to be connected with the steel frame with self-tapping screws through a connecting plate that is pre-welded on the I-steel beam by the workman.

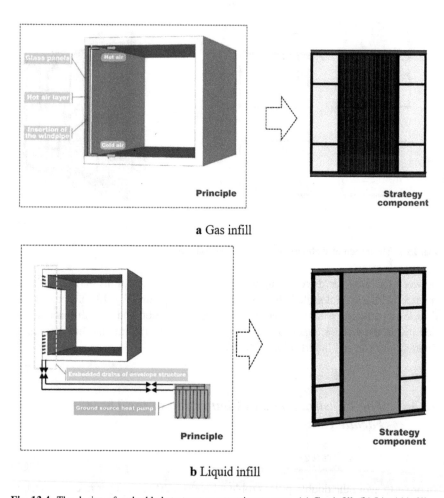

a Gas infill

b Liquid infill

Fig. 13.4 The design of embedded strategy construction process. (**a**) Gas infill. (**b**) Liquid infill

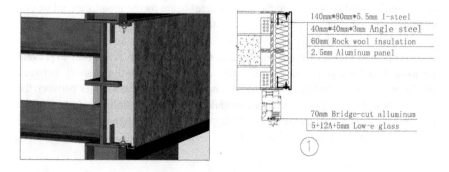

Fig. 13.5 Detail model and profile of cavity strategy

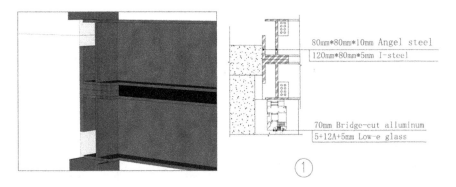

Fig. 13.6 Detail model and profile of shaft box

13.3.2.2 Shaft Box Strategy Type

Shaft box strategy module will also have a double-pillar problem in the connection with each other, so the same use of adjacent modules between the C-type steel column buckle plug-in will be adapted, with bolts by pre-welded on the outside of C-type steel nuts for connection and fixation. The I-steel of the outer connecting part of the upper and lower facade unit is covered with aluminum plate and rock wool insulation to solve the problem of hot and cold bridge. The inner side is connected to the main structure by angled steel, as shown in Fig. 13.6 . The gypsum board of the side partition of the module unit is connected to the I-steel frame with self-tapping screws through the connecting pieces welded on the I-steel beam in advance. The upper and lower horizontal bezels are connected in the same way as the side partitions.

13.3.2.3 Embedded Strategy Type

The facade unit of the embedded design strategy is a whole prefabricated wall panel consisting of a middle solid wall of 1500 mm * 3000 mm and two 750 mm * 3000 mm glass panels on both sides, with a casement window in the middle of the glass panel. The design strategy is divided into gas cycle embedded design strategy and liquid circulation embedded design strategy according to the fluid form introduced in the facade unit.

The embedded design strategy of gas circulation utilizes the form of XPS sandwich insulation of the light steel construction system of which the inside is fixed with self-tapping holding to 15 mm thick gypsum board with waterproof coating. The drains which use to storage heat are welded by each single vertical galvanized steel pipe and then welded with the horizontal tinned steel pipe of the reserved hole, forming the integral galvanized steel pipe which exhausts to the outside. The exhaust is fixed to the outside of the light steel structure by the self-tapping screw, and the two sides of the heat storage drains are also connected to the broken bridge

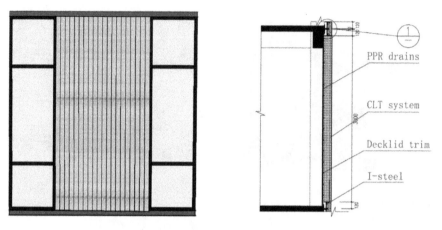

Fig. 13.7 Model and profile of water embed box strategy in CLT

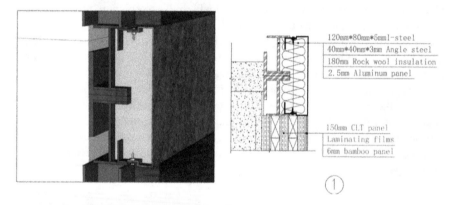

Fig. 13.8 Detail model and profile of embed strategy in CLT

aluminum window frame with the self-tapping screws. The outside of the heat storage drains is a double-layer hollow glass, with a reserved 30 mm air inter-layer to the drains to enhance the capacity of the heat storage pipe. The glass bridge aluminum window frame connected to light steel frame through self-tapping screws. Liquid-embedded facade unit and construction is shown in Figs. 13.7 and 13.8.

13.4 Construction Detail Design and Performance Simulation

The simulation software named Design Builder was used in this study to simulate the energy consumption of the design model and to verify the performance of three integrated prefabricated enclosure systems under different climatic conditions.

13.4.1 Cavity Strategy

13.4.1.1 Modeling

For the cavity strategy model with cavity depth of 600 mm, a reference model is set as a comparison, and the energy saving effects of facade walls with different structures and cavity strategy models were simulated to test. The information of the cavity strategy models and reference model is shown in Table 13.1.

13.4.1.2 Parameter Settings

In this study, three representative cities, namely, Beijing, Nanjing, and Guangzhou, will be selected as the research objects. The climate characteristics of the three cities are shown in Table 13.2. The building enclosure materials as well as the heat transmission coefficient of cavity strategy models in three wall structures and the reference model are shown in Table 13.3, and the opening time of the north and south windows of the basic reference model and the inner and outer windows of the cavity strategy models are shown in Table 13.4. During the daytime of winter, get outer window closed and the inner window open, so as to make the air in the cavity heated by radiation from the sun and then transmitted into the room by convection. During the

Table 13.1 Model information of cavity strategy and reference model

	Model.	An overview of the model
The basic reference models		Model width, depth, height dimensions: 3000 mmm * 6000 mm * 3000 mm South window: area ratio of window to wall:50%, window area 50% North window: window area of 100% Structure of external walls: basic wall structure applied to 4 facade walls
Cavity strategy		Model width, depth, height dimensions: 3000 mmm * 6000 mm * 3000 m Position of facade unit: South facade Cavity depth: 600 mm Outer window of cavity: area ratio of window to wall:100% window area: 100% inner window of cavity: area ratio of window to wall:50% window area 50%; North Window: window area 100% three south facade wall structures to be compared: light-gauge steel stud, PC, and CLT East/west/north facade wall: Basic wall structure

Table 13.2 Climatic characteristics of Beijing, Nanjing, and Guangzhou

City	Climate zone	Temperature characteristics	Latitude
Beijing	Cold zone	Average temperature of the coldest month = 0 ~ 10 °C	39° 56 'n
Nanjing	Hot summer and cold winter zone	The average temperature of the coldest month is 0 ~ 10 °C The average temperature of the hottest month is 25 ~ 30 °C	31°14'n
Guangzhou	Hot summer and warm winter zone	Average temperature of the coldest month >10 °C The average temperature of the hottest month is 25 ~ 29 °C	22°26'n

Table 13.3 Wall materials and heat transmission coefficient of cavity strategy models and reference model

The name of the enclosure structure	Enclosed structural materials	Heat transmission coefficient W/(m².K)
The basic wall structure	Beijing: 200 mm reinforced concrete + 40 mm XPS	0.553
	Nanjing: 200 mm reinforced concrete + 30 mm thick XPS	0.694
	Guangzhou: 200 mm reinforced concrete + 20 mm XPS	0.994
Light-gauge steel stud	6 mm bamboo planks + 15 mm air inter-layer + 15 mm HIP + 15 mm OSB + 80 mm XPS + 15 mm gypsum board	0.146
PC system wall	20 mm XPS + 40M concrete + 80 mm XPS + 80 mm concrete	0.224
CLT system wall	6 mm bamboo plank + 200 mm CLT	0.538
Ground	Project ground floor	0.264
Roof	Project flat roof	0.259
Window	5 + 12A + 5 double layer low-E insulating glass	1.628

Table 13.4 Opening time of Windows of cavity strategy models and reference model

Facade unit		Opening time
Cavity strategy Models	Outer window of cavity	All day long from May to September
	Inner window of cavity	January–April 9:00–17:00 May–September 19:00–8:00 October-December 9:00–17:00
	North window	May–September 19:00–8:00
The basic reference models	South window	May–September 9:00–17:00
	North window	May -September 9:00–17:00

nighttime of winter, the inner and outer windows are closed to provide a good effect to heat preservation. During the daytime of summer, the outer windows are open and the inner windows are closed, while the inner and outer windows are open at night in summer for natural ventilation. Windows in Nanjing and Guangzhou use a way of external shading and the opening time are in the same length with the cooling time. In the basic reference model, natural ventilation is used in the transition season, with north-south window closed in the heating and cooling season. The indoor temperature is set at 18 °C–26 °C, and the outdoor temperature is naturally ventilated within 18 °C–26 °C. The heating time is set from November to March, the cooling time is from May to September, and the window is also set to be closed at weekend.

13.4.1.3 Simulation Results and Analysis

The simulation results of the cavity strategies in three structures in Beijing, Nanjing, and Guangzhou are shown in Fig. 13.9. The simulation results data are the comparison of heating energy consumption, cooling energy consumption, and total energy

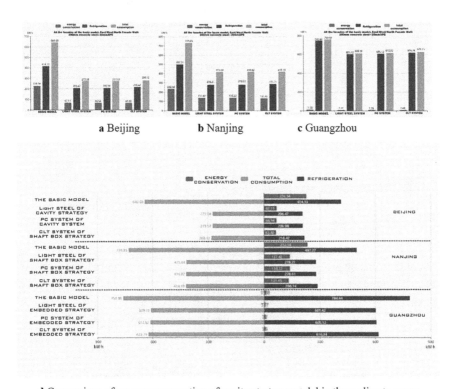

a Beijing   b Nanjing   c Guangzhou

d Comparison of energy consumption of cavity strategy model in three climate zones

Fig. 13.9 Summary of energy consumption comparison of well strategy. (**a**) Beijing. (**b**) Nanjing. (**c**) Guangzhou. (**d**) Comparison of energy consumption of cavity strategy model in three climate zones

consumption throughout the year in the model. Through comparative analysis, it can be found that the annual energy consumption of the basic model in Beijing area is 35.59 kW.h/m², and the consumption of light-gauge steel stud model, precast concrete system model, and CLT system model is 15.20 kW.h/m², 15.19 kW.h/m², and 15.56 kW.h/m². The cavity strategy models of three structures compared to the basic model save 56.98% of energy consumption on average, of which the precast concrete system model has the lowest annual energy consumption, saving 57.3%. The annual energy consumption of the basic model in Nanjing area is 40.54 kW.h/m², and the consumption of light-gauge steel stud model, precast concrete system model, and CLT system model is 23.15 kW.h/m² and 23.23 kW.h/m², respectively. The cavity strategy models of three structures save 42.9% of energy consumption on average compared with the basic model, of which the light-gauge steel stud model has the lowest annual energy consumption, saving 43.11%. The annual energy consumption of the basic model in Guangzhou area is 42.22 kW.h/m², and the consumption of light-gauge steel stud model, precast concrete system model, and CLT system model is 33.84 kW.h/m², 34.04 kW.h/m², and 34.65 kW.h/m². The cavity strategy models of three structures compared to the basic model save 19.06% of energy consumption on average, of which the light-gauge steel stud system model has the lowest annual energy consumption, saving 19.8%.

13.4.1.4 Conclusion

The cavity strategy models of three structures in Beijing compared to the basic model save 56.98% of energy consumption on average, 42.9% of energy consumption in Nanjing, and 19.06% in Guangzhou.

13.4.2 Shaft Box Design Strategy

13.4.2.1 Modeling

For the shaft box strategy model with a depth of 600 mm, a reference model is set as a comparison, and the energy-saving effects of facade walls with different structures and strategy models were simulated to test. The information of the shaft box strategy models and reference model is shown in Table 13.5.

13.4.2.2 Parameter Settings

The shaft box strategy of the facade unit of different construction systems and the building enclosing structure materials of the basic reference model, as well as the heat transmission coefficient as shown in Table 13.5, the opening time of the north-south window of the basic reference model and the inner and outer windows of the

Table 13.5 Model information of shaft box strategy and reference model

	Model	An overview of the model
The basic reference models		Model width, depth, height dimensions: 3000 mmm * 6000 mm * 3000 mm South window: area ratio of window to wall:50% window area 50% North window: window area of 100% Structure of external walls: basic wall structure applied to 4 facade walls
Shaft box Strategy		Model width, depth, height dimensions: 3000 mmm * 6000 mm * 3000 m position of facade unit: South facade; Shaft box depth: 150 mm outer window of shaft box: area ratio of window to wall:70% window area: 100% Size of louvre ventilation vent: 2700 mmm * 300 mm * 2 inner window of shaft box: area ratio of window to wall:50% window area 50% North Window: window area 100% Three south facade wall structures to be compared: light-gauge steel stud, PC and CLT East/west/north facade wall: Basic wall structure

shaft box strategy model is shown in Table 13.6, and the design of the winter day adopts the mode of closing the outer window and vents and opening the inner window, so that the heat in the cavity is quickly transmitted into the room through solar radiation. The building enclosure materials as well as the heat transmission coefficient of shaft box strategy models in different structures and the reference model are shown in Table 13.6, and the opening time of the north and south windows of the basic reference model and the inner and outer windows of the shaft box strategy models are shown in Table 13.7. During the daytime of winter, get outer window and vent closed and open the inner window, so as to make the air in the shaft box heated by radiation from the sun and then transmitted into the room by convection.

13.4.2.3 Simulation Results and Analysis

The simulation results of the shaft box strategy of different structures in Beijing, Nanjing, and Guangzhou are shown in Fig. 13.10. The simulation results data are the comparison of heating energy consumption, cooling energy consumption and, total energy consumption throughout the year in the model. Through comparative analysis, it can be found that the annual energy consumption of the basic model in Beijing area is 35.59 kW.h/m^2, and the consumption of light-gauge steel stud model,

Table 13.6 Wall materials and heat conductivity of shaft box strategy and reference model

The name of the enclosure structure	Enclosed structural materials	Heat transmission coefficient W/(m²) K)
The basic wall structures	Beijing: 200 mm reinforced concrete + 40 mm XPS	0.553
	Nanjing: 200 mm reinforced concrete + 30 mm thick XPS	0.694
	Guangzhou: 200 mm reinforced concrete + 20 mm XPS	0.994
Light-gauge steel stud	6 mm bamboo planks + 15 mm air inter-layer + 15 mm HIP + 15 mm OSB + 80 mm XPS + 15 mm gypsum board	0.146
Prefabricated concrete system walls	20 mm XPS + 40M concrete + 80 mm XPS + 80 mm thick concrete	0.224
CLT system wall	6 mm bamboo plank + 200 mm CLT	0.538
Ground	Project ground floor	0.264
Roof	Project flat roof	0.259
Window	5 + 12A + 5 Double layer Low-E insulating glass	1.628

Table 13.7 Opening time of Windows of shaft box strategy and reference model

Facade unit		Turn on the time setting
Shaft box strategy models	Blind vents	All day from May to September
	The outer window of the shaft box	April, October 9:00–17:00 May–September 19:00-8:00
Shaft box strategy models	The inner window of the shaft box	January–April 9:00–17:00 May–September 19:00–8:00 October–December 9:00–17:00
	North window	May–September 9:00–17:00
The basic reference model	South window	May–September 9:00–17:00
	North window	May–September 9:00–17:00

PC system, and CLT system model is 19.22 kW.h/m², 19.2 kW.h/m², and 19.16 kW.h/ m². The shaft box strategy models of three structures compared to the basic model save 45.99% of energy consumption on average, of which the CLT system model has the lowest annual energy consumption, saving 46.16%. The annual energy consumption of the basic model in Nanjing area is 40.54 kW.h/m², and the consumption of light-gauge steel stud model, PC system, and CLT system model is 26.76 kW.h/ m², 26.92 kW.h/m², and 26.75 kW.h/m². The shaft box strategy models of three structures save 33.8% of energy consumption on average compared with the basic model, of which the CLT system has the lowest annual energy consumption, saving 34.0%. The annual energy consumption of the basic model in Guangzhou area is 42.22 kW.h/m², and the consumption of light-gauge steel stud model, PC system, and CLT system model is 27.92 kW.h/m², 27.90 kW.h/m², and 28.08 kW.h/m². The

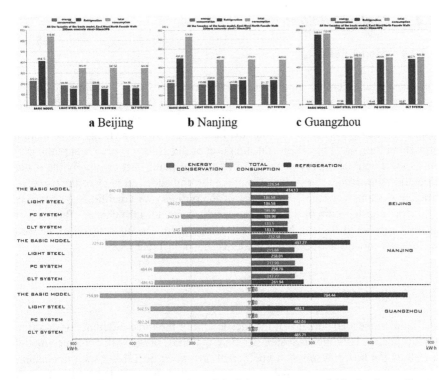

a Beijing   **b** Nanjing   **c** Guangzhou

d Comparison of energy consumption of shaft box strategy model in the three climate

Fig. 13.10 Summary of energy consumption comparison of shaft box strategy. (**a**) Beijing. (**b**) Nanjing. (**c**) Guangzhou. (**d**) Comparison of energy consumption of shaft box strategy model in the three climate zones

shaft box strategy models of three structures save 33.8% of energy consumption on average compared with the basic model, of which the PC system has the lowest annual energy consumption, saving 33.91%.

13.4.2.4   Conclusion

The shaft box strategy models of three structures in Beijing compared to the basic model save 45.99% of energy consumption on average, 33.8% of energy consumption in Nanjing, and 33.8% in Guangzhou.

13.4.3 Embedded Strategy

13.4.3.1 Model Building

Considering the scientific simulation, embedded strategy only uses liquid to do the computer simulation verification. The liquid embedded strategy of the facade unit is to set up a heat source drain system in the wall which is combined with the ground source heat pump system to simulate the heating and cooling energy consumption throughout the year. The energy consumption of the ground source heat pump system is the electrical energy generated by the heat pump that causes by the water circulation throughout the system. Compare the embedded strategy of the two construction systems with the basic model of the ordinary air-conditioning system in the energy consumption of the whole year which includes the energy efficiency of the three regions of Beijing, Nanjing, and Guangzhou. The information of the well box strategy models and reference model is shown in Table 13.8.

13.4.3.2 Parameter Settings

The liquid embedded strategy of the facade unit of the three different construction systems and the building enclosure structure materials of the basic reference model as well as the heat transmission coefficient are shown in Table 13.9, the opening time of the north and south windows of the basic reference model and the internal and external windows of the liquid embedded strategy model are shown in

Table 13.8 Model information of water embed strategy and reference model

	Model	An overview of the model
The basic reference model		Model width, depth, height dimensions: 3000 mmm * 6000 mm * 3000 mm South window: area ratio of window to wall:50% window area 50% North window: window area of 100% Structure of external walls: basic wall structure applied to 4 facade walls
Embedded strategy		Model width, depth, height dimensions: 3000 mmm * 6000 mm * 3000 m South window: area ratio of window to wall:50% window area 50% North Window: window area 100%; three south facade wall structures to be compared: light-gauge steel stud, PC and CLT East/west/north facade wall: Basic wall structure

Table 13.9 Wall materials and heat conductivity of water embed strategy and reference model

The name of the enclosure structure	Enclosed structural materials	Heat transmission coefficient W/(m²⁾ K)
The basic wall structures	Beijing: 200 mm reinforced concrete + 40 mm XPS	0.553
	Nanjing: 200 mm reinforced concrete + 30 mm thick XPS	0.694
	Guangzhou: 200 mm reinforced concrete + 20 mm XPS	0.994
Prefabricated concrete system walls	20 mm XPS + 40M concrete + 80 mm XPS + 80 mm thick concrete	0.224
CLT system wall	6 mm bamboo plank + 200 mm CLT	0.538
Ground	Project ground floor	0.264
Roof	Project flat roof	0.259
Window	5 + 12A + 5 Double layer Low-E insulating glass	1.628

Table 13.10 Opening time of Windows of water embedded strategy and reference model

Facade unit		Turn on the time setting
The basic reference models	South window	May–September 9:00–17:00
	North window	May–September 9:00–17:00
Liquid-embedded strategy models	South window	May–September 9:00–17:00
	North window	May–September 9:00–17:00

Table 13.10, and the basic reference model and the liquid-embedded strategy are both using excessive seasonal natural ventilation, heating, and cooling season with the closed north and south windows.

13.4.3.3   Simulation Results and Analysis

The simulation results of liquid embedded strategy of different structures in Beijing, Nanjing, and Guangzhou are shown in Figs. 13.11 and 13.12. The simulation results data are the comparison of heating energy consumption, cooling energy consumption, and total energy consumption throughout the year in the model. Through comparative analysis, it can be found that the annual energy consumption of the basic model in Beijing area is 35.59 kW.h/m², the light-gauge steel framing system is 9.46 kW.h/m², and the PC system is 9.51 kW.h/m², CLT system 9.66 kW.h/m². The liquid embedded strategy models of three construction systems saves an average of 73.2% of energy consumption compared with the basic model of which the light-gauge steel framing system has the lowest energy consumption throughout the year, saving 73.4% of the energy. The annual energy consumption of the basic model in Nanjing area is 40.54 kW.h/m², light-gauge steel framing system is 13.50 kW.h/m², PC system is 13.60 kW.h/m², CLT system is 13.70 kW.h/m². The liquid embedded

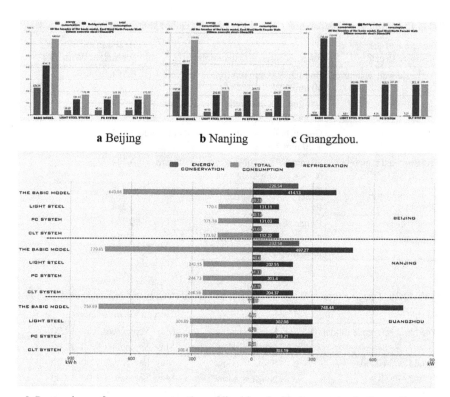

a Beijing   **b** Nanjing   **c** Guangzhou.

d Comparison of energy consumption of liquid embedded strategies in three climate zones.

Fig. 13.11 Summary of energy consumer comparison of embedded strategy. (**a**) Beijing. (**b**) Nanjing. (**c**) Guangzhou. (**d**) Comparison of energy consumption of liquid embedded strategies in three climate zones

strategy models of three construction systems saves an average of 66.4% of energy consumption compared with the basic model, of which the PC system has the lowest energy consumption throughout the year, saving 66.6% of the energy. The annual energy consumption of the basic model in Nanjing area is 42.22 kW.h/m², light-gauge steel framing system is 17.05 kW.h/m², PC system is 17.11 kW.h/m², and CLT system is 17.13 kW.h/m². The liquid-embedded strategy models of three construction systems saves an average of 59.5% of energy consumption compared with the basic model, of which the light-gauge steel framing system has the lowest energy consumption throughout the year, saving 56.6% of the energy.

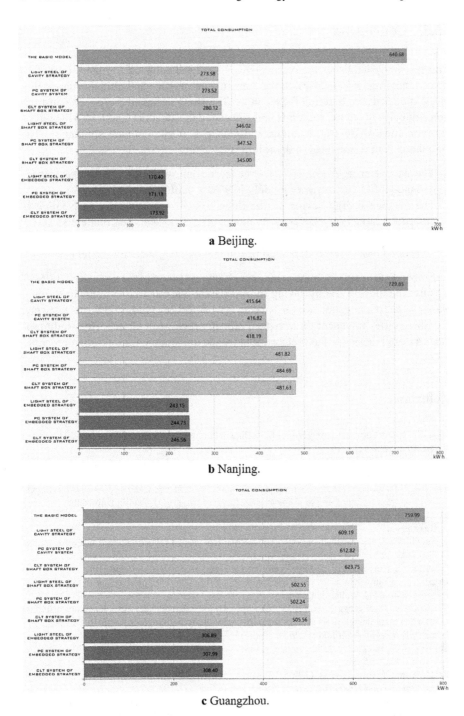

Fig. 13.12 Summary of energy comparison of three strategies in Beijing, Nanjing, and Guangzhou. (**a**) Beijing. (**b**) Nanjing. (**c**) Guangzhou

13.5 Conclusion

This research establishes the perspective on the sustainable performance of prefabricated buildings and focuses on the construction practice and climate adaptability of the prefabricated building facade units. The study designed 3 types of construction strategies of cavity, shaft box and embedded facade units, and tested the physical properties of Beijing, Nanjing, and Guangzhou by computer simulation. The simulation data results lead to the following conclusion:

- The average energy saving of the three facade unit strategies in Beijing is embedded model (73.1%) > cavity model (56.9%) > shaft box model (45.9%).
- The average energy saving of the three facade unit strategies in Nanjing is embedded model (66.3%) > cavity model (42.9%) > shaft box model (33.8%).
- The average energy saving of the three facade unit strategies in Guangzhou is embedded model (59.0%) > shaft box model (33.8%) > cavity model (19.6%).

Embedded strategy of light-gauge steel framing system in the three design strategies has the highest energy saving in of different construction systems in Beijing, Nanjing, and Guangzhou which is up to 73.1%, 66.3%, and 59.0%, respectively. The three cities represent three typical climatic regions in China. This study provides a design basis for the prefabricated facade units in the future design process.

References

Chen, X. (2011). Eames House: The art of dwelling in the context of industrialization [J]. *Huazhong Architecture, 29*(12), 16–19.

Crosbie, M. (1997). *The passive solar design and construction handbook* (pp. 16–18). Steven Winter Associates.

Guo, J. (2013). *Research on the integration performance of industrialized building envelop components of affordable housing in the severe cold zone of China [D]*. Tianjin University.

Haller, F., & Wichmann, H. (1989). *System-design*, Fritz Haller:Bauten, Möbel, Forschung[M]. Birkhäuser.

Hovestadt, V., & Hovestadt, L. (1999). The ARMILLA project [J]. *Automation in Construction, 8*(3), 325–337.

Kieran, S., & Timberlake, J. (2003). *Refabricating architecture [M]*. McGraw-Hill.

Knaack, U., Chung-Klatte, S., & Hasselbach, R. (2012). *Prefabricated systems: Principles of construction[M]*. Walter de Gruyter.

Konstantinou, T., & Knaack, U. (2013). An approach to integrate energy efficiency upgrade into refurbishment design process, applied in two case-study buildings in Northern European climate [J]. *Energy & Buildings, 59*(4), 301–309.

Lin, H. (2003). Integrality design and analysis of sustainable architecture system [J]. *Architecture Journal*, (12), 15–17.

Lin, Z., Song, Y., Sun, J., et al. (2015). Climate responsive modular building skin design strategy [J]. *Eco-city and Green Building, 2*, 54–61.

Lin, Z., Song, Y., & Han, D. (2017). Tectonics, space and performance: Three dimensionalities for the development of lightweight prefabricated architecture from 20th century [J]. *Urbanism and Architecture, 13*, 12–16.

Ni, L. (2010). *Research on "more with less" eco-design thoughts and application [D]*. Shanghai Jiaotong University.

Song, Y. (2000). *Integrated design with nature-research on ecological architectural design [M]*. China Construction Industry Press.

Standard for assessment of prefabricated building, GB/T 51129-2017, Ministry of Construction of the People's Republic of China.

Zhu, N. (2013). *Tectonic with technic: On elements, system and cases of building process from the perspective of manufacturing [D]*. Tsinghua University.

Zimmermann, M. (2009). *IEA ECBCS ANNEX 50 prefabricated systems for low energy renovation of residential buildings*. Jahresbericht 2008[R]. Swiss: EMPA.

Chapter 14
Architectural Design Evaluation: Recent Reform of the Singapore Green Mark to Prompt a Hypothesized Revolution of the Green Building Rating System

Yifan Song, Stephen Siu-Yu Lau, and Siu-Kit Lau

14.1 Introduction

Green building movement has been around for almost 50 years. During this period, there has always been a very obvious phenomenon, that is, the war between architects and the field of green building. Many architects protested that design innovation was replaced by green function (Alter, 2009, 2010; Cilento, 2010). Almost most architects acquiesced in the fact that architectural design was limited by green design practices (Rebecca, 2015). Contrary to architects, green advocates criticized the impractical and wasteful design (Capps, 2009) and claimed that caring about the environment and architectural design are not contradictory (Hosey, 2012; Ahuja, 2012). It seems that architecture and "green" do not conflict theoretically. However, the correct theory does not mean that it is feasible in practices. The green building market, so far, is still subject to the concept of profit. As Windapo (2014) proposed, economy is the main driver of green building rather than environment. Therefore, stakeholders like investors or building owners tend to maximize short-term benefits with minimal investment. Due to such output-based purpose of stakeholders, credit responsibilities of green building rating systems (GBRSs) are assigned to the project design teams that include architects, and then everyone starts to design with the mission of achieving the requirements of criteria, which can be called "design for criteria." For that reason, the GBRS, as a key driving force of the green building movement, should be explored for their validity and effectiveness.

Y. Song (✉) · S.-K. Lau
School of Design and Environment, National University of Singapore, Singapore, Singapore
e-mail: songyifan@u.nus.edu; slau@nus.edu.sg

S. S.-Y. Lau
Department of Architecture, The University of Hong Kong, Hong Kong, China

14.1.1 Analysis of GBRS's Potential Problems from a Historical Perspective

In order to get the reasons for the war between architects and "green," it is essential to be traced back to the mid-nineteenth century – the origin of the environmental movement. At that time, there was an increasing notification of environmental problems caused by rapid industrialization (1760–1840). In the building sector, many architectural worked firstly responded to environmental issues, such as Antoni Gaudí's Casa Mila (1912), Le Corbusier's Villa Savoye (1931), Frank Lloyd Wright's Fallingwater (1937) and Alvar Aalto's Villa Mairea (1939). However, these early modernist practices drew the public's attention more for their design styles than their environmentally responsive considerations. Even their so-called environmentally responsive design was not really "green." For example, Casa Mila just borrowed natural organic forms in the building envelope and interior design. Le Corbusier explained his designs by biological analogies, but his practice such as Villa Savoye didn't show the actual connection between architecture and nature very well. In 1962 Silent Spring written by Rachel Carson made the environmental movement initiate in academia. Thereafter, Victor Olgyay, as a pioneer of bioclimatism, proposed a new term "ecological building" from the perspective of architecture in 1963. However, "ecological building" did not become the mainstream of building environmental movement, since two oil crises (1973 and 1979) derived a more attractive term "green building movement." In this movement, people's focus was moving toward energy and resource conservation that required modern technology to support. Therefore, the engineer with technological and mechanical knowledge became the main driver for the development of GBRSs, while the architect did not participate. Under this circumstance, almost all developed and developing GBRSs had characteristics related to energy, resource, modern technology, and mechanism, e.g., an energy-based framework, elements/components of construction, and quantifiable credits. This performance-oriented and component-based evaluation method based on Analytic Hierarchy Process violates the nature of architectural design. Through architectural design, architects create a "realm," a complete coexistence system that connects history, culture, society, climate, nature, environment, and people, instead of a simple building enframed[1] by components. Although it is difficult for GBRS to consider such unmeasurable and indescribable architectural design, this does not mean that it is not important and should not be incorporated into GBRSs.

---

[1] The verb enframed is the past singular of enframe. It is the derivative of the German word Ge-siell proposed by Heidegger and Lovitt (2013). Here its connotation is gathering, i.e. assembling and ordering, things through a summons forth from criteria.

14.1.2 The Latest Researches About Architectural Design in the Green Building Movement

Based on the review of the literature on GBRSs in the past decade, the research contents related to architectural design are extracted and then summarized into four aspects. Such four aspects illustrate the importance of architectural design. Firstly, the regional context was emphasized. Neama (2012) explored that the international GBRS – Leadership in Energy and Environmental Design (LEED) – was not applicable to some parts of the world. Yamany et al. (2016) further studied the applicability and implementation of LEED in local buildings and found that the inapplicable aspects include culture, society, environment, climate, and economy. Even though many efforts were exerted so as to develop LEED to fit into other contexts, it was difficult for GBRSs to cover every aspect of different regional contexts, especially the unmeasurable aspect such as culture and society. For architects, the most damaging effects of design could be the failure to embrace place and enhance local identity (Hosey, 2012). Secondly, the evaluation of green buildings should go beyond the scope of building and consider larger scales such as site, neighborhood, or urban scale. Almost all GBRSs were constructed into five criteria "Site," "Energy," "Resource," "Indoor Environment Quality" (IEQ), "Innovation." A large number of researches have explored the effectiveness in "Energy," "Resource," and "IQE" except "Site." Conte and Monno (2012) believed that the sustainability of buildings was not only about the building itself, but also its resilience in the urban matrix, so they proposed a cross-scale evaluation model to complement GBRSs. Huo et al. (2017) analyzed the site aspect by comparing several GBRSs and also pointed out that site aspect required the comprehensive coordination by architects and site planners to ensure that a site is functionally efficient, aesthetically pleasing, and environmentally sustainable. Breaking through the building scale is not simply an enlargement of the scale; it also requires the overall planning ability of architectural design. Thirdly, passive design was proven to have significant energy saving benefits. Several researchers found that the use of passive design like natural ventilation, air cooling, and shade provision reduces the considerable percentage of energy consumption (Taleb, 2014; Kamal, 2012). However, current GBRSs are so specialized in helping to improve energy efficiency and reduce the resource consumption that it neglects to guide the architect toward real solutions to achieve passive design. Fourthly, user's behavior and wellbeing have become the new frontier of green building movement, which is a revolutionary reform to make the public no longer only focus on energy and resource. With the long-term use of green buildings, a large number of studies have used building operation data and post occupancy evaluation to prove that user's awareness and behavior directly determined the actual "green" performance of buildings (Zhao et al., 2015; Scofield, 2016; Mamalougka, 2013). Other studies have learned through questionnaire surveys and interviews that people in green buildings are not comfortable or satisfied and even fall into sick building syndrome (Gou et al., 2013; Altomonte & Schiavon, 2013). There is a missing link between green buildings and their users (Deuble & de Dear, 2012). In summary, regional context, large-scale coordination, passive design, and user-orientation are main parts of conventional architectural design. Now that they have raised issues in green building movement, the GBRS, as a key driving force, need to take them into account. Hence, it is imperative to incorporate architectural design into GBRSs.

14.2 Comparative Analysis of Architectural Design Evaluation in Singapore Green Mark

Singapore Green Mark (GM) was the first GBRS of the tropics. Many characteristics of its old version.[2] were very similar to LEED until it was overhauled in 2015. The latest version GM NRB v2015 updates the traditional energy-based framework of GBRSs into a new framework that consists of five criteria "Climatic Responsive Design," "Building Energy Performance," "Resource Stewardship," "Smart and Healthy Building," and "Advanced Green Efforts" with the same weighting in order to balance the importance of all aspects (Table 14.1). It aims to promote a greater recognition to climatically contextual design and tropical vernacular and to propose the accessible architectural design guide for green benchmarking by new criterion "Climatic Responsive Design," and to provide users with a high-quality experience from a spatial scale by new criterion "Smart and Healthy Building," which indicates the formal incorporation of architectural design concepts into GM NRB.

This research will compare all versions of Green Mark for Non-Residential Buildings (GM NRB) launched in the past 10 years. They are GM NRB v4.1 launched in 2010, GM NRB v2015 and GM NRB v2015 R.[3] Sub-criteria and credits related to architectural design will be identified and analyzed to study how Singapore incorporates the architectural design into the GBRS.

Table 14.1 Framework reform of GM NRB

Old criteria	GM NRB v4.0	New criteria	GM NRB 2015
Energy Efficiency	116 (61.05%)	Climatic Responsive Design	30 (21.43%)
		Building Energy Performance	30 (21.43%)
Water Efficiency	17 (8.95%)	Resource Stewardship	30 (21.43%)
Environmental Protection	42 (22.11%)		
Indoor Environmental Quality (IEQ)	8 (4.21%)	Smart and Healthy Building	30 (21.43%)
Other Green Features	7 (3.68%)	Advanced Green Efforts	20 (14.28%)

---

[2] Old version of Green Mark refers to any version launched before 2015

[3] GM NRB 2015 R includes three modifications of GM NRB v2015 before November 1, 2019.

Table 14.2 Comparison of credits in the sub-criterion "Sustainable Urbanism"

Old credits	GM NRB v4.0	New credits		GM NRB v2015	GM NRB v2015 R
–	–	Bicycle Parking		P	P
–	–	Environmental Analysis		2	2
–	–	Environmental Analysis -Ecosystems		1*	1*
Green Transport	4	Response to Site Context		3	3
		Green Transport	Parking Infrastructure	1.5	1.5
			Bicycle Usage Promotion Design	0	0.5
–	–	Urban Heat Island (UHI) Mitigation		1	1
Maximum Scores (Weighting)	4 (2.11%)	Maximum Scores (Weighting)		5 + 1* (4.29%)	5 + 1* (4.29%)

14.2.1   Comparison of Site and Landscape

According to Table 14.2, GM NRB v4.0 only involves "Green Transport." It was also investigated by Huo et al. (2017) who pointed out that GM has the lowest weighting compared to other GBRSs. To make up for this flaw, "Sustainable Urbanism" in GM NRB v2015 is the sub-criteria with the largest increase in new credits (Table 14.2). New credits "Environmental Analysis," "Response to Site Context," and "Urban Heat Island (UHI) Mitigation" respectively set up new requirements about comprehensive analysis of the surrounding environment before site design, comprehensive considerations of site design, and design materials, which is in line with the thinking way of conventional architectural design.

Since green transport needs both large-scale coordination and small-scale specific settings, the content of old credit "Green Transport" is divided into two new credits. The contents about site access and connectivity are moved to "Response to Site Context" to work with new added aspects such as urban context, site topography and hydrology, site micro climate, etc., to achieve a comprehensive site design. The contents about facilities and equipment continue to be classified in the new "Green Transport." In GM NRB v2015 R, design checklist is added to new credit "Green Transport – Bicycle Usage Promotion Design" in order to guide designers to "design for human." In summary, new sub-criteria "Sustainable Urbanism" try to take into account the nature of architectural design such as larger scales, site design, comprehensive consideration, and human-centered design.

As a garden city, Singapore has mature landscape standards, e.g., Landscape Excellence Assessment Framework (LEAF), Active Beautiful Clean Waters Design Guidelines (ABC Water), Landscaping for Urban Spaces and High-Rises (LUSH) program, etc.

In GM NRB, landscape aspect is not updated much (Table 14.3). Almost all new credits continue to use the same name and evaluation content of old credits. Their update is only to increase the difficulty of evaluation through decreasing the score

Table 14.3 Comparison of credits in the sub-criterion "Integrated Landscape and Waterscape" and related credits in the criterion "Resource Stewardship"

Old credits	GM NRB v4.0	New credits			GM NRB v2015	GM NRB v2015 R
Green Plot Ratio (GnPR)	6	Green Plot Ratio (GnPR)			3	3
	–	GnPR $\geq$ 5.0			1*	1*
Tree Conservation	1	Tree Conservation (TC)	Preservation of Existing trees		0.5	0.5
			Replanting Native Species		0.5	0.5
–	–	Sustainable Landscape Management (SLM)	50% Native Species	LEAF Certification	0.5	0.5
–	–		Full scores of Environmental Analysis		0.5	0.5
Composts from Horticultural Waste	1		Landscape Management Plan		0.5	0.5
Stormwater Management	3	Sustainable Stormwater Management (SSM)	Infiltration Design	ABC Certification	1	1
Maximum Scores (Weighting)	11 (5.79%)	Maximum Scores (Weighting)			5 + 1* (4.29%)	5 + 1* (4.29%)
Drought Tolerant Plant	1	Drought Tolerant Plant			0.5	0.5
Rainwater Harvesting	1	Rainwater Harvesting			1	1
Maximum Scores (Weighting)	2 (1.05%)	Maximum Scores (Weighting)			1.5 (1.07%)	1.5 (1.07%)

of "Greenery Provision," adding new advanced item "GnPR $\geq$ 5.0," splitting the credit "Tree Conservation," and introducing standard certification (ABC Water) in "Sustainable Stormwater Management." The new credit "Sustainable Landscape Management" introduces LEAF certification and adds the requirements for 50% native planting and more landscape management measures. Both of which are also extracted from credits in LEAF. With emerging the evaluation of native planting, the proportion requirement of "Drought Tolerant Plant" is decreased from 80% to 20%. In addition, the full scores under "Environmental Analysis" in Table 14.2 can also get an additional 0.5 point here. Because the two referenced standards have corresponding design guidelines, GM NRB does not consider the architectural design too much in recent reform.

Table 14.4 Comparison of credits in the sub-criterion "Tropicality"

Old credits	GM NRB v4.0	New credits			GM NRB v2015	GM NRB v2015 R
–	–	Air Tightness and Leakage			P	P
Envelope Thermal Transfer Value	12	Envelope and Roof Thermal Transfer			P	P
Envelope Design/ Thermal Parameters	5					
	30	Tropical Façade Performance	Checklist	Simulation	3↑	3
	0	Internal Spatial Organization (ISO)	Locating Non-Thermally Critical Spaces		1	1↑
Ventilation in Common Area	5		Ventilation Mode of Transient Common Spaces		2↑	2
Vertical Greenery	1	Vertical Greenery			1*↓	1*
–	–	Thermal Bridging			1*	1*
–	–	Low Heat Gain Façade			–	1*
Natural Ventilation— Opening Toward Prevailing Wind Directions	10	Ventilation Performance (VP)	Checklist: Opening Toward Prevailing Wind Directions	Simulation of Wind Velocity/ Thermal Comfort/Air Quality	4	4
–	–		Checklist: Depth of Room vs Opening			
Natural Ventilation— Simulation of Wind Velocity	10	–				
–	–	Wind Driven Rain Simulation			1*	1*
Maximum Scores (Weighting)	Null	Maximum Scores (Weighting)			10 + 3* (9.29%)	10 + 4* (10.00%)

14.2.2   Comparison of Passive Design and Human-Centered Design

For the evaluation of passive design (Table 14.4), the old credit "Envelope Design / Thermal Parameters" are assigned to pre-requisite "Envelope and Roof Thermal Transfer," credit "Tropical Façade Performance," and new credit "Internal Spatial Organization" in order to distinguish overall thermal parameters, detailed thermal parameters, and space-related content.

In "Envelope and Roof Thermal Transfer," the Envelope Thermal Transfer Value (ETTV) and Roof Thermal Transfer Value (RTTV) are combined for evaluation. In

"Tropical Façade Performance," the evaluation object becomes "all walls" instead of "west facing wall" in the old version. Two new thermal parameters, Glass Shading Coefficient and Skylight U-value, is added in order to minimize thermal heat gain while daylighting. There are also new prerequisite "Air Tightness and Leakage," advanced items "Thermal Bridging," and "Low Heat Gain Façade." These adjustments for thermal parameters make the evaluation of building envelope more integral and comprehensive only on the technical level. The only one that has guidance on architectural design is Window-to-Wall Ratio (WWR) in "Tropical Façade Performance."

GM NRB also realizes that passive design cannot only make requirements on the technical level, so it adds a new credit "Internal Spatial Organization" to guide architects from the design level. A subtle note (that is why there is a 0 point in the Table 14.4) associated with spatial design in old credit "Envelope Design / Thermal Parameters" is upgraded to a new credit "Internal Spatial Organization – Locating Non-Thermally Critical Spaces" and further updated the percentage requirement in GM NRB v2015 R to achieve the quantitative purpose. Besides, the score of natural ventilation in "Ventilation Mode" is increased from 1.5 to 2 to emphasize the importance of passiveness.

In addition, "Natural Ventilation" is renamed to "Ventilation Performance" that adds not only the new design strategy "Depth of Room vs Opening" but the simulation of thermal comfort, air quality, and wind driven rain as well. In "Depth of Room vs Opening," single-sided ventilation, cross ventilation, design for spatial scale, and atrium design are specifically mentioned.

Therefore, the sub-criteria "Tropicality" interprets that shaping building passive design should consider its façade, interior layout, orientation, opening, as well as space type that all belong to the architectural design. Last but not least, the biggest change in this part from the new version to the old version is not in the above analysis but in the framework. Just as the previous analysis, the framework of old version determines that the passive design can be overlapped by active design (Song & Lau,

Table 14.5 Comparison of credits in the sub-criterion "Spatial Quality"

Old credits	GM NRB v4.0	New credits		GM NRB v2015	GM NRB v2015 R
Daylighting—Common Areas	3	Lighting	Daylighting—Common Areas	2	2↑
Daylighting—Occupied Areas	3		Daylighting—Occupied Areas	4 *New*	4↑
Artificial Lighting	12		Artificial Lighting	1 *New*	1
Noise Level	1	Acoustics		2 *New*	2
Innovation—Roof Garden	1*	Wellbeing	Biophilic Design	3	3
–	–				
–	–		Universal Design (UD) Mark	1	1
Maximum Scores (Weighting)	20 (10.53%)	Maximum Scores (Weighting)		10 (7.14%)	10 (7.14%)

2019). That is why there is the "Null" in the Table 14.4. New framework establishment with new criterion "Climatic Responsive Design" contributes to the equality between passive design and active design.

As can be seen from Table 14.5, the sub-criteria "Spatial Quality" aims to create human access to quality daylight, make space acoustically comfortable, and evoke a connection to nature. The evaluation content of "Daylighting-Occupied Areas," "Artificial Lighting," and "Acoustics" is brand new, although their name of credits can correspond to the old ones. Therefore, their content cannot be compared horizontally and can only be compared through the scores. In "Lighting," the score ratio of "Daylighting -Common Areas," "Daylighting-Occupied Areas," and "Artificial Lighting" change from 1:1:4 to 2:4:1, which means that the focus of lighting shifts from active design to passive design. However, all these contents are performance evaluation in which there is no architectural design guidance. In GM NRB v2015 R, "Daylighting-Common Areas" adds automatic lighting controls while "Daylighting-Occupied Areas" adds the evaluation mitigation strategies such as blinds controlled with daylight sensors, variable opacity glazing, bi-level glazing, and fitted glazing, in order to further improve the lighting quality. For "Wellbeing", both "Biophilic Design" and "Universal Design Mark" are architectural design credits. "Biophilic Design" includes accessible sky gardens, indoor planting, biomimicry designs, natural forms or ecological attachment, and images of nature, which is more biased toward landscape design and interior design. The "Universal Design Mark" includes no detailed evaluation content but directly requires the certification of local standards.

14.2.3 Summary

After identifying the relevant sub-criteria of architectural design, the study calculates the percentage of them in GM NRB v2015 R, of which "Passive Design" accounts for the largest percentage (10%), followed by "Human-Centered Design" (7.14%) (Fig. 14.1). Architectural design credits are further identified in these criteria (Table 14.6). It can be found that 81.8% of the credits (27 of the 33 credits) are design credits of architectural design, while the rest 18.2% credits are technical credits of architectural design.

Fig. 14.1 Weightings of four sub-criteria

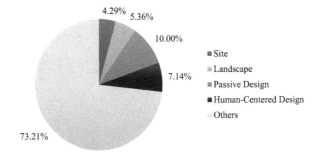

Table 14.6 Architectural design credits of GM NRB v2015 R

	No.	Credits	New	Design/technical credit	Evaluation method				Scores of design checklist vs simulation
					Checklist			Simulation method	
					Qualitative credit	Quantitative credit	Reference design standard		
Site	1	Environmental Analysis	✓	D	✓	□	□	✓	1:2
	2	Environmental Analysis—Ecosystems	✓	D	□	□	□	✓	□
	3	Response to Site Context	✓	D	✓	□	□	✓	1:3
	4	Urban Heat Island (UHI) Mitigation	✓	D	✓	✓	□	□	□
	5	Bicycle Parking	✓	D	□	✓	✓	□	□
	6	Green Transport—Parking Infrastructure	□	D	□	✓	✓	□	□
	7	Green Transport—Bicycle Usage Promotion Design	✓	D	✓	□	□	□	□

	No.	Credits	New	Design/technical credit	Evaluation method			Simulation method	Scores of design checklist vs simulation
					Checklist		Reference design standard		
					Qualitative credit	Quantitative credit			
Landscape	8	Green Plot Ratio (GnPR)	□	D	□	✓	✓	□	□
	9	TC—Preservation of Existing trees	□	D	✓	□	□	□	□
	10	TC—Replanting Native Species	□	D	✓	□	□	□	□
	11	SLM—50% Native Species	✓	D	✓	✓	✓(LEAF)	□	□
	12	SLM—Full scores of Environmental Analysis	✓	D	□	□	□	✓	□
	13	SLM—Landscape Management Plan	✓	D	✓	□	□	□	□
	14	SSM—Infiltration Design	□	D	✓	□	✓(ABC)	□	□
	15	Drought Tolerant Plant	□	D	✓	□	□	□	□
	16	Rainwater harvesting	□	D	✓	□	□	□	□

(continued)

Table 14.6 (continued)

Passive Design	No.	Credits	New	Design/technical credit	Evaluation method Checklist Qualitative credit	Quantitative credit	Reference design standard	Simulation method	Scores of design checklist vs simulation
	17	Air Tightness and Leakage	✓	T	□	□	✓ (SS212 & 381)	□	□
	18	Envelope and Roof Thermal Transfer	□	T	□	✓	✓ (BCA Code)	□	□
	19	Tropical Façade Performance	□	T	□	✓	□	✓	3:3
	20	Thermal Bridging	✓	T	□	□	✓ (SGBC)	□	□
	21	Low Heat Gain Façade	✓	T	□	✓	✓ (BCA)	□	□
	22	Vertical Greenery	□	D	✓	✓	□	□	□
	23	ISO—Locating Non-Thermally Critical Spaces	✓	D	✓	✓	□	□	□
	24	ISO—Ventilation Mode of Transient Common Spaces	□	D	✓	✓	□	□	□
	25	VP—Opening towards Prevailing Wind Directions	□	D	✓	✓	□	✓	3:4
	26	VP—Depth of Room vs Opening	✓	D	✓	✓	□	□	
	27	Wind Driven Rain Simulation	✓	D	□	□	□	✓	□

	No.	Credits	New	Design/technical credit	Evaluation method				Scores of design checklist vs simulation
					Checklist				
					Qualitative credit	Quantitative credit	Reference design standard	Simulation method	
Human-Centered Design	28	Daylighting—Common Areas	□	D	✓	□	□	□	□
	29	Daylighting—Occupied Areas	✓	D	✓	□	□	✓	4:4
	30	Artificial Lighting	✓	T	□	✓	✓	□	□
	31	Acoustics	✓	D	✓	✓	□	□	□
	32	Biophilic Design	✓	D	✓	□	□	□	□
	33	Universal Design Mark	✓	D	□	□	✓ (UD)	□	□

GM NRB makes effort to incorporate architectural design into it in all aspects to a large extent. On the one hand, architectural design is considered from the criteria "Climatic Responsive Design" to four sub-criteria and finally to 33 credits. On the other hand, in order to make these architectural design credits gradeable, GM NRB tries its best to quantify design credits of architectural design. As can be seen from Table 14.6, 8 credits quantify qualitative content; 6 credits directly quote the quantitative data or calculation methods from relevant standards; 5 credits admit standard certification, e.g., LEAF, ABC,UD, etc.; 2 credits only use simulation methods (strictly speaking, these 2 credits are architectural design credits without strategic guidance); 6 credits add simulation methods as a parallel option; and 8 credits are still qualitative.

14.3 Discussion from the Perspective of Architecture

14.3.1 Continuity of Architectural Design Process

Table 14.6 shows that there are six technical credits of architectural design. "Air Tightness and Leakage," "Envelope and Roof Thermal Transfer," "Tropical Façade Performance," "Thermal Bridging," and "Low Heat Gain Façade" are technical credits about walls, and "Artificial Lighting" is technical credits about luminaire, which all belong to the phases of "Developed Design" or "Technical Design" in the architectural design process (RIBA, 2013). The new version of GM NRB put them in confusion with design credits in the same sub-criteria mainly because of the Analytic Hierarchy Process (AHP) decision-making method and the performance fundamentals of GBRSs. The AHP Method disassembles the GBRSs into "Site," "Energy," "Resource," "IEQ," and "Innovation" according to environmental classification and then continues to break the building into "elements of construction," i.e., components, for performance evaluation. Such hierarchical classification completely ignores the continuity of architectural design process. As a result, due to higher weightings, the work of "Developed Design" or "Technical Design" may be moved from the middle and the late cycles of architectural design process to the very front of it, which disrupts conventional design thinking, thus reducing the quality of architect's concept design. Therefore, the GBRSs should consider the continuity of credits and the connection between design credits and technical credits.

14.3.2 Integrity of Architectural Design

As described in the introduction, performance-oriented and component-based GBRSs have been developed for 30 years since BREEAM launched in 1990. The component-level evaluation of "Energy," "Resource," and "IEQ" becomes more integral. Take the technical credit "Tropical Façade Performance" as an example. It can be found that with continuous reforms, the components involved in building envelope have been added as much as possible and coordinated under one credit, which makes the evaluation of building envelope more integral and comprehensive. However, most GBRSs have not yet realized the integrity of architectural design evaluation, perhaps because architectural design has just begun to be focused in recent 10 years, or architectural design itself is too unmeasurable and indescribable to be incorporated into very quantitative rating systems. Fortunately, GM NRB v2015 has taken a step forward in this regard. The establishment of new sub-criteria "Sustainable Urbanism," "Internal Spatial Organization," and "Spatial Quality" shows that GM NRB has realized that the so-called integrity of architectural design lies in the site and space. "Sustainable Urbanism" is hardly affected by old credits, so its evaluation is relatively integral. But "Internal Spatial Organization" and "Spatial Quality," affected by the framework, characteristics and evaluation content of old versions, are not integral. The "WWR" in "Tropical Façade Performance" and the "Opening Towards Prevailing Wind Directions," and "Depth of Room vs Opening" in "Ventilation Performance" are all related to space design, and both are design credits. Hence, integrating thermal, ventilation, light, and acoustics into one sub-criteria or credit can help architects to comprehensively design based on the spatial scale.

14.3.3   Interactivity During Architectural Design

Simulation in GBRSs is a very common evaluation method to quantify the environmental impact of architectural design credits. It is a kind of approximate imitation in which specific data is obtained by rehearsing the evaluated object through software. It often occurs in middle or late design phases. In other words, an ideal relationship, theoretically, is that simulation serves as a scientific method to optimize the architect's comprehensive design. There is a harmonious cooperation between architectural design and simulation. Only through going back and forth between design and validation can excellent design be produced. But the status quo in GM NRB v2015 is not so. There are six credits that set design checklist and simulation as two parallel options instead of requiring both evaluation methods to be performed, and what is worse is that simulation's scores are greater than or equal to design's scores. Driven by the "design for criteria," stakeholders have a high probability of choosing the simulation method that has a higher score and can directly export specific data reports. Simulation is then moved from late design phase to the very early phase. Thus, green design becomes a more quantitative and measurable simulation-driven and criteria-driven design, which is the result of software analysis by importing values of the factors evaluated in GBRSs. The architect's comprehensive consideration of other aspects, especially qualitative and unmeasurable aspects such as large-scale culture and history or small-scale human behavior, cannot be integrated into this process. Even so, simulation is still liked by clients, because simulation-driven design drastically lowers the time it takes for design teams to develop a design and its generated report can be used to get more scores. Finally, clients always bias toward simulation in green projects, which will intensify the war between architects and "green." Therefore, GBRSs should carefully consider the relationship between the design checklist and simulation and their scores.

14.4   Conclusion

Due to the environmental movement and energy crisis, "green" began with energy, resource, and modern technology, which makes its rating system more technological, mechanical, performance-based, and component-oriented. There has been a lack of connection between "green" and "architecture" for 50 years until an increasing number of studies have discovered the importance of architectural design in the past decade. However, environmental sophistication in GBRSs and architectural design sophistication in practice don't blend well, so that it's difficult for GBRSs to incorporate architectural design into GBRSs. With the vision of supporting architectural design, Singapore Green Mark was overhauled in GM NRB v2015 which subvert the traditional GBRS's framework. It allocates a total weight of 26.79% to the architectural design in which 81.8% credits are design credits and 18.2% credits are technical credits. Four new sub-criteria about site, landscape, passive design, and human-centered design make effort to involve architectural design strategies, such as large-scale design, local context, facade, opening, orientation, spatial layout, space depth, space type, natural ventilation design strategy, material selection, etc. Four evaluation methods are applied to make them more measurable. Although GM NRB v2015 strives to strengthen architectural design in various aspects of evaluation framework, evaluation content, and evaluation methods, there are still some key flaws that violate the nature of architectural design, thus invalidating the relevant credits. Because architectural design focuses on process, comprehensive consideration, and cooperation with others, GBRSs should consider the continuity of architectural design process, integrity of architectural design, and interactivity during architectural design at the standard level, thereby driving architects to participate in "green" in practice.

References

Ahuja, R. (2012). Lean and green construction. *International Journal of Scientific & Engineering Research, 3*(7), 1–4.
Alter, L. (2009). *Why is so much green architecture so ugly?* Treehugger.
Alter, L. (2010). *Frank Gehry starts architectural bunfight with comments on LEED and green building.* Treehugger.
Altomonte, S., & Schiavon, S. (2013). Occupant satisfaction in LEED and non-LEED certified buildings. *Building and Environment, 68,* 66.

Capps, K. (2009). Green building blues. *American Prospect*.

Cilento, K. (2010). Gehry vs LEED. *ArchDaily*.

Conte, E., & Monno, V. (2012). Beyond the buildingcentric approach: A vision for an integrated evaluation of sustainable buildings. *Environmental Impact Assessment Review, 34*, 31–40.

Deuble, M. P., & De Dear, R. J. (2012). Green occupants for green buildings: The missing link? *Building and Environment, 56*, 21–27.

Gou, Z., Prasad, D., & Lau, S. S.-Y. (2013). Are green buildings more satisfactory and comfortable? *Habitat International, 39*, 156–161.

Heidegger, M., & Lovitt, W. (2013). *The question concerning technology, and other essays*. HarperCollins Publishers.

Hosey, L. (2012). *The shape of green: Aesthetics, ecology, and design*. Island Press.

Huo, X., Yu, A. T. W., & Wu, Z. (2017). A comparative analysis of site planning and design among green building rating tools. *Journal of Cleaner Production, 147*, 352–359.

Kamal, M. A. (2012). An overview of passive cooling techniques in buildings: Design concepts and architectural interventions. *Acta Technica Napocensis: Civil Engineering & Architecture, 55*, 84–97.

Mamalougka, A. (2013). *The relationship between user satisfaction and sustainable building performance*. Construction Management & Engineering (CME) Master, Delft University of Technology.

Neama, W. A. S. A. (2012). Protect the planet through sustainability rating systems with local environmental criteria – LEED in the Middle East. *Procedia - Social and Behavioral Sciences, 68*, 752–766.

Rebecca. (2015). *The aesthetics of green design*. Available from: http://www.sophersparn.com/the-aesthetics-of-green-design/

RIBA. (2013). *RIBA plan of work 2013 template*. Royal Institute of British Architects.

Scofield, J. H. (2016). *Do green buildings really save energy? A look at the facts* [Online]. GreenBiz. Available: https://www.greenbiz.com/article/do-green-buildings-really-save-energy-look-facts [Accessed].

Song, Y., & Lau, S. S. Y. (2019). Connecting theory and practice: An overview of the natural ventilation standards and design strategies for non-residential buildings in Singapore. *International Review for Spatial Planning and Sustainable Development, 7*, 81–96.

Taleb, H. M. (2014). Using passive cooling strategies to improve thermal performance and reduce energy consumption of residential buildings in U.A.E. buildings. *Frontiers of Architectural Research, 3*, 154–165.

Windapo, A. (2014). Examination of green building drivers in the South African construction industry: Economics versus ecology. *Sustainability, 6*, 6088–6106.

Yamany, S. E., Afifi, M., & Hassan, A. (2016). Applicability and implementation of U.S. Green Building Council Rating System (LEED) in Egypt (A longitudinal study for Egyptian LEED certified buildings). *Procedia Environmental Sciences, 34*, 594–604.

Zhao, D.-X., He, B.-J., Johnson, C., & Mou, B. (2015). Social problems of green buildings: From the humanistic needs to social acceptance. *Renewable and Sustainable Energy Reviews, 51*, 1594–1609.

Chapter 15
Architectural Design for Manufacturing and Assembly for Sustainability

Vikrom Laovisutthichai and Weisheng Lu

15.1 Introduction

"Sustainable development," or in other words "sustainability," can be defined as development that meets the needs of the present without compromising the ability of future generations to meet their own needs (Brundtland et al., 1987). It concerns the dynamic relationship between the human economic system and greater ecological mechanism, in which humankind can continuously develop and flourish without jeopardizing the diversity, complexity, and function of the ecosystem (Costanza, 1992). It is supported by all United Nations (UN) members through the establishment of 2030 agenda for sustainable development with 17 Sustainable Development Goals (SDGs) (UN, 2015). The SDGs are the new global targets, replacing Millennium Development Goals in 2000–2015 (Sachs, 2012). The SDGs urge every country and organization to collaborate together for the global well-being of the current and next generations by deliberating three broad categories of challenges: economic development, environmental sustainability, and social inclusion.

The SDGs have been repetitively analyzed, reinterpreted, and detailed into different sectors and contexts. Sustainability has become a doctrine of our social economic activities. The Architecture, Engineering, and Construction (AEC) industry, being one of the major pollution and waste generators, also adopts this doctrine to

V. Laovisutthichai (✉)
Department of Real Estate and Construction, The University of Hong Kong, Pok Fu Lam, Hong Kong

Department of Architecture, Chulalongkorn University, Bangkok, Thailand
e-mail: vikrom@connect.hku.hk

W. Lu
Department of Real Estate and Construction, The University of Hong Kong, Pok Fu Lam, Hong Kong

S. S. Y. Lau et al. (eds.), *Design and Technological Applications in Sustainable Architecture*, Strategies for Sustainability, https://doi.org/10.1007/978-3-030-80034-5_15

minimize the negative impacts it can impose. Many efforts, such as the sustainable development framework for implementation (Hill & Bowen, 1997), fiscal incentives and regulations for a higher level of sustainable development (Pitt et al., 2009), and education and training for construction professionals (Shi et al., 2014), have been made. Sitting at the beginning of a building's life cycle, architectural design, if meticulously considered, can provide a great opportunity in supporting sustainability (Ding, 2008). Partly due to this opportunity, the Royal Institute of British Architects (RIBA) contextualizes 17 SDGs in architectural design, identifies eight sustainable outcomes with evaluation methods, and publishes the guidelines for real-life practice (RIBA, 2019).

To achieve the SDGs, numerous design thinkings have been proposed, e.g., a passive design, green building material selection, building envelope insulation, energy consumption minimization, modular design, adaptive reuse of an existing structure, and design for energy-efficient deconstruction and recycling (Akadiri et al., 2012; Sassi, 2006). The open or adaptable building is also recommended as a preparation for future changes in an organization, occupants, lifestyles, households, and technology (Kendall, 1999). Moreover, the new construction and computational technologies, e.g., Building Information Model (BIM), can also facilitate the architectural design process for sustainable architecture (Wong & Fan, 2013). These sustainable architecture design thinkings firstly concentrate on the design techniques to reduce energy consumption during the building use (Ding, 2008; Owen & Dovey, 2008). They were demanded extension, by realizing that other stages of building life cycle, namely, construction, renovation, and demolition, also consume a large number of resources and affect the building sustainability performance. Any design consideration that can help improve broader environmental and social aspects should also be investigated and implemented.

Design for Manufacturing and Assembly (DfMA) is arising as one of the most promising design thinkings for achieving sustainable architecture design and construction (Wasim et al., 2020). DfMA is generally a design philosophy and methodology with special attention given to manufacturing and assembly to facilitate construction flow, reduce cost, shorten the time, prevent potential waste, and ultimately increase efficiency (Gao et al., 2020). Although DfMA has considered some sustainability factors (e.g., waste minimization), the two have not been consciously discussed together for achieving broader SDGs. Moreover, numerous DfMA design suggestions and instructions are proposed in various sources, and there is currently a lack of integrated process for designers (Lu et al., 2020; Tan et al., 2020). Its advantages in terms of sustainability also remain nebulous. These drawbacks make the implementation of DfMA a wearisome and laborious task to search, gather, and scrutinize information before utilizing it into a design. What is urgently needed is an integrated process for DfMA to achieve SDGs in real-life practice.

This chapter, therefore, aims to provide an integrated DfMA process and identify its sustainable outcomes. It is achieved by reviewing literature and conducting case studies. The remainder of this chapter is structured as follows. Section 15.2 offers the research methods adopted. Section 15.3 displays the background of DfMA. It is followed by the integrated DfMA process in Sect. 15.4 and its benefits to sustainability in Sect. 15.5. Finally, it reaches the discussion and conclusion parts.

15.2   Research Methods

This research comprises four stages (see Fig. 15.1). It began with the data collection from the search on the Google Scholar search engine, using the keywords, e.g., Design for Manufacturing and Assembly, Design for Construction, and DfMA. The second screening process was completed manually to confirm the relevance to the main research questions. The selected research should be written in English, focused on the construction context, and related to (a) DfMA definitions and principles, (b) practice examples, and (c) application advantages. Then, the DfMA recommendations and examples from various sources were amalgamated and aligned with the RIBA plan of work (RIBA, 2020) to generate the tentative DfMA process.

The tentative process is validated and improved by two case studies. The first empirical case is the design and construction of Modular Integrated Construction (MiC) Exhibition Center developed by the Hong Kong Construction Industry Council (CIC) in Kowloon Bay, Hong Kong SAR. This building is the first full volumetric prefabrication building in Hong Kong and the resource center to provide information for the industry stakeholders. It is also one of the pilot projects to forecast the future of modular construction in Hong Kong and alleviate the construction industry challenges, e.g., the shortage of construction labor, high construction cost, and high demand in construction service. This two-story building consists of ten modules and one prefabricated fire service installation (FSI) unit. The author visited

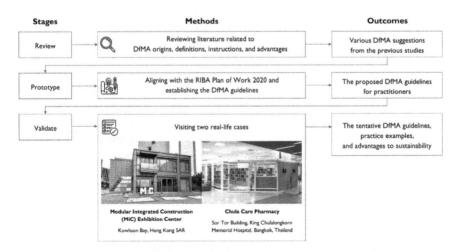

Fig. 15.1 Research methods

the case and talked to the CIC staff to gain more in-depth information about the real-life design practice and results.

The second case is the interior design and renovation of Chula Care drugstore developed by King Chulalongkorn Memorial Hospital in Bangkok, Thailand. This 120 square meter pharmacy shop is located at the lobby of the Sor Tor Building, surrounded by the Outpatient Department (OPD), patients' waiting area, information kiosk, and office. The design and construction must be completed while the healthcare service is operated as usual. To reduce noise and adverse effects on the patients and medical staffs, DfMA and prefabrication are adopted. The primary author actively participated in this project as one of the design team members to gain the insight and lessons learned from real-life practice.

Finally, the collected data from both cases were used for the DfMA process validation and improvement. To identify the benefits of DfMA to sustainability, the results from the implementation were also matched with the RIBA sustainable outcomes (RIBA, 2019) and SDGs (UN, 2015). These actions were taken to encourage DfMA and ultimately position this design approach in the design of sustainable architecture.

15.3 Design for Manufacturing and Assembly: Origin and Definition

DfMA is originated in the manufacturing industry, since the recognition that design, if carefully practiced, can present a valuable opportunity to simplify manufacturing and assembly processes and diminish the total cost significantly. By adopting DfMA, "the wall" between design and manufacturing and assembly is destroyed (Boothroyd, 1994). This effective approach is established under the umbrella of Design for Excellence (DfX), where "X" or Excellence can be replaced by many terms for different sectors, project characteristics, and purposes, such as safety, waste reduction, reliability, and operations (Kuo et al., 2001). The specific aspects of DfX include Design for Environment (DfE), Design for Reliability (DfR), Design for Disassembly (DfD), Design for Construction Waste Minimization (DfCWM) (Laovisutthichai et al., 2020), inter alia.

With lacklustre productivity, the construction industry also borrows DfMA to mitigate barriers and escalate productivity in every construction method (Gao et al., 2020; Tan et al., 2020). Many efforts have been made for migrating the concept in the AEC industry, e.g., the critical review of DfMA in construction (Gao et al., 2020), identification of factors influencing DfMA adoption in Singapore (Gao et al., 2018), comprehensive report for practitioners (BCA, 2017), and practice example in the UK construction (Banks et al., 2018). Tan et al. (2020) also create a soft-landing for DfMA in construction by establishing practicable guidelines from the real-world construction case. Literally, DfMA is not entirely new to the industry, as some principles are similar to the previous concepts such as lean construction, value management, and buildability (Lu et al., 2020).

In construction, DfMA is defined as "both a philosophy and a methodology whereby products are designed in a way that is as amenable as possible for downstream manufacturing and assembly" (Gao et al., 2020). DfMA helps the construction process by the optimization of building components, joints, and processes without jeopardizing the design and its functionality (Lu et al., 2020). This approach challenges the heterogeneity, discontinuity, and disintegration characteristics in the industry and intends to break down the silos among AEC stakeholders (Chen & Lu, 2018). Its application requires the amendment in professional practice from traditional to non-linear collaborative design method (Pasquire & Connolly, 2003).

15.4 DfMA Process

With the vogue of DfMA, numerous design instructions and recommendations are provided in different sources and various disciplines. By reviewing previous literature comprehensively, 11 DfMA recommendations are found. To help its implementation, they need to be fit into the established architecture design process rather than advocating an entirely new process. We notice that the RIBA Plan of Work (RIBA, 2020) is an established framework for the architecture design process. Some DfMA works have been added to the Plan of Work as an extra layer. Likewise, we adopt the similar approach and collect, analyze, assemble, and align DfMA for sustainability on the RIBA Plan of Work's process, which can be perceived from four design stages: (1) preparation and briefing; (2) conception; (3) spatial coordination, and (4) technical design. The tentative working process is then validated through two case studies in Hong Kong SAR and Thailand (see Fig. 15.2).

Stage 1 Preparation and Briefing
According to the RIBA (2020) Plan of Work, the design process starts with receiving the design brief from a client, including the project goals, requirements, budget, location, and intended outcomes. Then, designers collect more data, broaden their perspective, redefine, and prepare the required information for the concept development in the following process. To achieve DfMA, several actions should be taken at the beginning, as described follows.

- Multidisciplinary team, including the project owner, designer, contractor, and manufacturer, is needed to be built from the early stage. As DfMA requires additional information related to manufacturing and assembly processes, the engagement of other stakeholders can help the information flow during design (Gerth et al., 2013). Many studies and reports, including the MiC Exhibition Center, agree that this practice is one of the keys to success (AIA, 2019; Banks et al., 2018; Maas & van Eekelen, 2004). In the case of façade design (Chen & Lu, 2018), for example, the multidisciplinary team could share information and give valuable suggestions to the design immediately. It significantly contributed to the increase in manufacturability of the curtain wall system and construction cost saving.

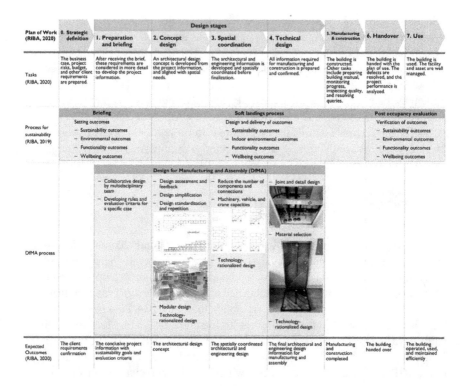

Plan of Work (RIBA, 2020)	0. Strategic definition	1. Preparation and briefing	2. Concept design	3. Spatial coordination	4. Technical design	5. Manufacturing & construction	6. Handover	7. Use
Tasks (RIBA, 2020)	The business case, project risks, budget, and other client requirements are prepared.	After receiving the brief, these requirements are considered in more detail to develop the project information.	An architectural design concept is developed from the project information, and aligned with spatial needs.	The architectural and engineering information is developed and spatially coordinated before finalization.	All information required for manufacturing and construction is prepared and confirmed.	The building is constructed. Other tasks include preparing building manual, monitoring progress, inspecting quality, and resolving queries.	The building is handed with the plan of use. The defects are resolved, and the project performance is analyzed.	The building is used. The facility and asset are well managed.
Process for sustainability (RIBA, 2019)	**Briefing** Setting outcomes – Sustainability outcomes – Environmental outcomes – Functionality outcomes – Wellbeing outcomes			**Soft landings process** Design and delivery of outcomes – Sustainability outcomes – Indoor environmental outcomes – Functionality outcomes – Wellbeing outcomes			**Post occupancy evaluation** Verification of outcomes – Sustainability outcomes – Environmental outcomes – Functionality outcomes – Wellbeing outcomes	
DfMA process		**Design for Manufacturing and Assembly (DfMA)** – Collaborative design by multidisciplinary team – Developing rules and evaluation criteria for a specific case	– Design assessment and feedback – Design simplification – Design standardization and repetition – Modular design – Technology-rationalized design	– Reduce the number of components and connections – Machinery, vehicle, and crane capacities – Technology-rationalized design	– Joint and detail design – Material selection – Technology-rationalized design			
Expected Outcomes (RIBA, 2020)	The client requirements confirmation	The conclusive project information with sustainability goals and evaluation criteria	The architectural design concept	The spatially coordinated architectural and engineering design	The final architectural and engineering design information for manufacturing and assembly	Manufacturing and construction completed	The building handed over	The building operated, used, and maintained efficiently

Fig. 15.2 DfMA process

- Rules and evaluation criteria for a specific case should also be established in the first stage. It can be utilized to weigh how design alternatives affect manufacturing, logistics, and assembly, make the design development straightforward, and ultimately increase the project constructability (Fox et al., 2001; Gerth et al., 2013). This strategy is proposed based on the nature of the AEC sector, which depends heavily on a specific project context, and there is no one size fit all assessment form for every project. In the drugstore case, as the construction site was adjacent to the OPD and patients' waiting area, the impact on these activities became one of the main criteria during design and construction.

Stage 2 Concept Design

In the second stage, the design team collaborates with other stakeholders to develop an architectural design concept, building mass, and conceptual planning based on the project information. Several DfMA instructions should be adopted in this stage.

- Design simplification is crucial when designing the building mass and form, as the complicated design means additional tasks and efforts during construction. In the MiC Exhibition Center, the module form is decided to be a simple rectangular shape. While in the bridge design (Kim et al., 2016), the form simplification became one of four main design principles. The components were simplified by rethinking the components' function and shape and interdependency with the other components. By doing so, it consumed fewer resources, reduced the number of the manufacturing process, and ultimately decreased the construction cost.

- Design standardization and repetition refer to the repetitive use of standardized components from the early design stage (Tam et al., 2007; Tan et al., 2020), since design variations make manufacturing and assembly process more complicated. It can enable mass production in construction, decreases construction cost, shortens time, and increases manufacturability, assemblability, and transportability of a project (Kurokawa, 2005). Many cases adopted this suggestion, e.g., a façade design (Arashpour et al., 2018) and residential tower design (Kurokawa, 1977). The drugstore case also used this strategy for the interior fixtures and shelves display design. Although the shelves' shape and dimensions were standardized, the store design was not wearisome and monotonous. The combination of five types of shelves, together with other design techniques, could make the space more appealing.
- Modular design is the design concerning the proportional relationship between the whole and its particular parts or components (Russell, 2012). In the drugstore design, the proportional relationship between the entire store area and shelves provided flexibility for modification. The cable-stayed solar tower design case reveals that its application led to a considerable reduction in cost, time, and risk during construction (Peterseim et al., 2016). This design technique can also encourage construction waste minimization during replacement, renovation, and demolition (Xu & Lu, 2019).
- Technology-rationalized design is also suggested to be utilized during concept design, spatial coordination, and detail design stages, as presently there are many construction and computational tools to assist practitioners, improve product quality, and reduce rework (Tan et al., 2020). For example, BIM is used to analyze the dimensional quality of prefabricated modules or components while design (Rausch et al., 2020). BIM is also recommended by BCA (2016), Kaner et al. (2008), and the case of MiC Exhibition Center, to assist the building information management during both design and construction.
- Design assessment and feedback from stakeholders are necessary for concept development (Meiling & Sandberg, 2009). In New York, apart from the designers' considerations, the conceptual design was sent to the manufacturer, transporter, and general contractor, to reconfirm the constructability and receive feedback before the detailed design (Ramaji et al., 2017).

Stage 3 Spatial Coordination
In the third stage, the architectural and engineering information is developed, assessed, and spatially coordinated before finalization. The design team works together with other parties to ensure the possibility of the design scheme. The building components, dimension, space arrangement, and other engineering systems must be considered in more detailed. Some DfMA suggestions can be used in this stage.

- The number of components and connections reduction is to simplify the construction procedure. In the façade design case, the team realized this strategy and selected the unitized curtain wall system since it required a smaller number of curtain wall components (Chen & Lu, 2018). The adoption of this method could simplify the process, decrease the risk of errors, and reduce the complexity of components' connections (Chen & Lu, 2018; Kim et al., 2016).
- Machinery, vehicle, and crane capacities should be pondered and planned from design to facilitate the flow and avoid errors when constructing (BCA, 2016). For instance, in the high-rise office building construction in the Netherlands, the weight of the building and components were designed following the capacity of a factory, trailer, and crane (Maas & van Eekelen, 2004). In the drugstore case, as the components were manufactured in the factory before transporting and placing on the site, the lightweight aluminum frames, shelves, and furniture were carefully designed with the consideration of pickup trucks' size to increase transportability and overall efficiency.

Stage 4 Technical Design

- The final design stage mainly focuses on detail design and project delivery. All building information is detailed, confirmed, documented, and submitted for manufacturing and construction. There are two DfMA suggestions to be implemented, as described below.
- Joint and detail are crucial parts of a building. It not only affects the building performance but also shapes the manufacturing, logistics, and assembly processes. In Malaysia, the multidisciplinary team designed the structural joint to be assembled safely, quickly, and easily (Jaganathan et al., 2013). The case of MiC Exhibition Center highlights that apart from the structural joint, the mechanical, electrical, and plumbing connections must also be well considered.
- Materials should be selected with the careful considerations of its origin, size, weight, specifications, and durability. The selection of lightweight, cost-effective, green, and durable material is also recommended to avoid material damage during construction and decrease maintenance cost in the long run (Gerth et al., 2013; Maas & van Eekelen, 2004). It is supported by the case study in China that lightweight steel-framed structure with thin-walled profiles was used. Although the structure's price was high, it could save a considerable amount of time and reduce risks during construction (Tan et al., 2020). Chen and Lu (2018) reveals that the use of sustainable, high fire resistance rate, and long lifespan materials as much as possible can improve the building durability and financial performance in the long run. The light-weight aluminium structure and standard materials were also selected for the drugstore design, because they required less on-site activities and facilitated transportation.

15.5 DfMA Sustainable Outcomes

Sustainable architecture design refers to materializing the built environment by adapting to local social economic, cultural, and environmental contexts, considering its consequences to the future generation (Guedes et al., 2009). The RIBA (2019) supports this doctrine, contextualizes the 17 SDGs in architectural design, and offers sustainable goals for practitioners. Eight RIBA sustainable outcomes include net-zero operational carbon, net-zero embodied carbon, sustainable water cycle, sustainable connectivity and transport, sustainable land use and ecology, good health and well-being, sustainable communities and social value, and sustainable life cycle cost. By comparing the outcomes of DfMA with these indicators, we reveal that the implementation of DfMA can increase sustainability in construction in many ways (see Fig. 15.3).

Outcome 1 Good Health and Well-Being

- *Workforce health and safety:* Several DfMA techniques, e.g., design standardization, simplification, and the number of components and connections reduction, support offsite construction, and decrease on-site process. In the previous research and two case studies, they help moving construction activities to a factory-like environment, offering a better working environment for construction workers, and reducing risks of many health and safety issues on-site, e.g., a falling of object or person from height (Banks et al., 2018; Jaganathan et al., 2013).

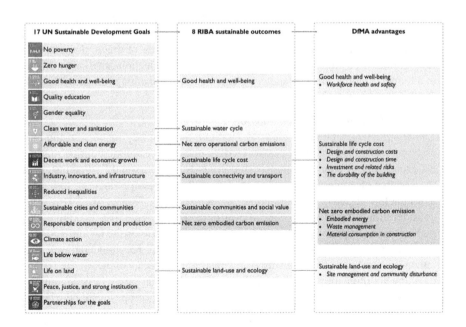

Fig. 15.3 DfMA sustainable outcomes

Outcome 2 Sustainable Life Cycle Cost

- *Design and construction costs:* The design complication and variation mean additional tasks and costs required in the manufacturing and assembly process. Some intricate design may need unique materials with extraordinary attachment process. DfMA recommendations support the simplification, repetition, and standardization of building components and form and contribute to a significant construction cost reduction (Chen & Lu, 2018; Kurokawa, 1977). It is also prospected to reduce the high cost of prefabrication construction markedly.

- *Design and construction time:* The implementation of DfMA decreases building design and construction duration in several ways. In the MiC Exhibition Center case, the joint and detail were designed with careful attention to simplify and shorten the component production and assembly processes. In the drugstore case, DfMA and prefabrication eliminate in situ works and significantly decrease the construction duration on site. The use of standardized material also reduces material subtraction time (Arashpour et al., 2018). Chen and Lu (2018) report that the implementation of DfMA in the curtain wall design could save the assembly time by 7 minutes per one component. For developers, the time reduction when applying DfMA means a considerable decrease in interest charges and early return of investment capital (KPMG, 2016).

- *Investment and related risks:* The careful consideration of construction process and coordination among stakeholders can ensure the manufacturability, transportability, assemblability, and feasibility of the design from the early stage (AIA, 2019; Fox et al., 2001; Maas & van Eekelen, 2004). DfMA also destroys barriers among parties, facilitates information flow, prevents risks of error and rework from misunderstanding, and ultimately increases sustainability performance.

- *The durability of the building:* DfMA prioritizes the use of green, durable, and low-maintenance building materials. These material selection criteria contribute to a longer building lifespan and economic benefits in the long run (Banks et al., 2018; Jaganathan et al., 2013).

Outcome 3 Net-Zero Embodied Carbon

- *Embodied energy:* DfMA intends to facilitate and simplify manufacturing and assembly tasks. This practice can also reduce energy consumption during materials processing, production, transportation, and on-site assembly. For instance, the number of components reduction contributes to less production, logistics, and attachment activities required (Chen & Lu, 2018; Kim et al., 2016). The consideration of vehicle size in the drugstore design could also maximize the pickup truck capacity and reduce the number of trips between the factory and site.

- *Material consumption in construction:* Several DfMA recommendations also lessen resources and materials used in construction. For example, in the bridge case, the simplification and standardization in design can save a large amount of material from design variations and reduce cost without compromising the building quality (Kim et al., 2016).

- *Waste management:* Apart from the material consumption, the early consideration of construction tasks makes the site and resources management more efficient and contributes to a well-managed construction site and waste (Gerth et al., 2013). Some DfMA instructions, e.g., careful material selection, dimensioning, and design standardization, also minimize cutting-to-fit waste, which is the predominant source of waste worldwide (Poon et al., 2004). In the drugstore, DfMA removed a large amount of potential waste on site and made the construction site and surroundings clean and safe.

Outcome 4 Sustainable Land Use and Ecology

- *Site management and community disturbance:* The removal of many construction activities from a site to controlled manufacturing line not only provides a safe working environment to construction workers but also significantly reduces the adverse effect on-site. In the MiC Exhibition Center case, DfMA and modular construction could reduce vibration, noise, and environmental impact on the surrounded community. The drugstore case also reveals that DfMA could support the interior renovation process without jeopardizing the patients' safety and interrupting the medical service around the site.

15.6 Discussion

DfMA as a Method for Sustainability

By reviewing literature and investigating two case studies, we affirm that DfMA is one of the design methods to achieve sustainability in the AEC industry like others, e.g., a passive design, durable and green material, reuse of existing structure, and open plan building. The implementation of 11 DfMA suggestions during four key design stages can contribute to four out of eight SDGs and RIBA sustainable outcomes: net-zero embodied carbon, sustainable land use and ecology, good health and well-being, and sustainable life cycle cost. This proactive approach encourages sustainable development, especially during the production and assembly stages. To endorse sustainability in the entire building life cycle, the use of DfMA, together with other sustainable architecture design methods, is highly recommended.

The Use of DfMA

This empirical study also substantiates that DfMA can be adopted by both architectural and interior design to increase the sustainability level in building construction and renovation. DfMA should be considered and included in a design from the beginning stage. However, the two case studies disclose that their utilization methods and final designs are varied. One DfMA suggestions must be selected carefully depending on specific project purposes and conditions, as it would be extremely difficult, if not entirely impossible, to include all of them into one design. Besides, the implementation of DfMA is not just to take something. Before including it into a design, these various suggestions need to be adjusted to suit the project requirements through the reinterpretation and adaptation processes.

Contributions and Limitations

This research constitutes both knowledge and practical contributions. It pieces together various DfMA notions, verifies its advantages to sustainability, and highlights one of the core challenges from the real-world application. For practitioners, it also makes DfMA easier for implementation by consolidating instructions from various sources, aligning with the RIBA Plan of Work 2020, generating integrated DfMA process, and providing real-life practice example. However, this research also has several limitations. Firstly, this research is conducted in the context of building construction in Hong Kong and interior renovation in Thailand. It still requires additional cases in real-world settings for generalization. Moreover, the proposed DfMA process and sustainable outcomes should be considered as the preliminary version. It should be further fine-tuned using more empirical studies and remains open to the new research and practice development in this discipline.

Future Studies

Future studies are recommended to explore other sustainable architecture design methods in real-life settings since this research approach can help researchers understand the practices and challenges profoundly. Although DfMA suggestions for architectural and interior design tend to be similar, this research discloses that their detail explanations and usages are different. Presently, many studies only concentrate on architectural design, while DfMA for interior architecture is rarely discussed in the previous literature. To fulfil this gap, additional studies are highly demanded. Moreover, this research reveals the complexity of DfMA interpretation and adaptation processes. To help its realization, more in-depth investigation and assistive tools should be further studied.

15.7 Conclusion

To achieve the UN sustainable development goals (SDGs) and RIBA sustainable outcomes, many architectural design thinkings have been proposed. Although Design for Manufacturing and Assembly (DfMA) has a high potential to help achieve these goals, DfMA and architectural design for sustainability are rarely discussed together in previous research or practice. Besides, the DfMA process itself and its sustainable outcomes remain unclear. To fill these knowledge voids, this research consolidates various DfMA practices, generates an integrated DfMA process, and verifies its advantages to achieving sustainability.

Through literature review and case studies, this chapter corroborates that DfMA is one of the proactive approaches to achieve the four SDGs and RIBA sustainable outcomes: net-zero embodied carbon, sustainable land use and ecology, good health and well-being, and sustainable life cycle cost. The implementation of 11 DfMA suggestions contributes to sustainability in construction and renovation. However, the use of DfMA is not just to take them. The two cases reveal that the implementations of DfMA suggestions during four design stages are varied, depending on the reinterpretation and adaptation to fit a particular project context.

This research amalgamates various guidance of DfMA, substantiates its abilities to realize SDGs, and uncovers the complexity behind its implementation from real-life case studies. However, the proposed DfMA process and results are merely the preliminary version, and require more investigation for generalization and improvement. Future research is suggested to explore other sustainable architecture design methods, interior design for manufacturing and assembly, and assistive tools to promote the utilization of DfMA.

References

Akadiri, P. O., Chinyio, E. A., & Olomolaiye, P. O. (2012). Design of a sustainable building: A conceptual framework for implementing sustainability in the building sector. *Buildings, 2*(2), 126–152.

American Institute of Architects (AIA). (2019). *Design for modular construction: An introduction for architects.*

Arashpour, M., Miletic, M., Williams, N., & Fang, Y. (2018). Design for manufacture and assembly in offsite construction: Advanced production of modular façade systems. In *ISARC. Proceedings of the international symposium on automation and robotics in construction* (Vol. 35, pp. 1–6). IAARC Publications.

Banks, C., Kotecha, R., Curtis, J., Dee, C., Pitt, N., & Papworth, R. (2018). Enhancing high-rise residential construction through design for manufacture and assembly–a UK case study. *Proceedings of the Institution of Civil Engineers-Management, Procurement and Law, 171*(4), 164–175.

Boothroyd, G. (1994). Product design for manufacture and assembly. *Computer-Aided Design, 26*(7), 505–520.

Brundtland, G. H., Khalid, M., Agnelli, S., Al-Athel, S., & Chidzero, B. J. N. Y. (1987). *Our common future* (Vol. 8). Oxford University Press.

Building and Construction Authority (BCA). (2016). *BIM for DfMA (Design for Manufacturing and Assembly) essential guide.* Author.

Building and Construction Authority (BCA). (2017). *Prefabricated prefinished volumetric construction.* Author.

Chen, K., & Lu, W. (2018). Design for manufacture and assembly oriented design approach to a curtain wall system: A case study of a commercial building in Wuhan, China. *Sustainability, 10*(7), 2211.

Costanza, R. (1992). *Ecological economics: The science and management of sustainability.* Columbia University Press.

Ding, G. K. (2008). Sustainable construction—The role of environmental assessment tools. *Journal of Environmental Management, 86*(3), 451–464.

Fox, S., Marsh, L., & Cockerham, G. (2001). Design for manufacture: A strategy for successful application to buildings. *Construction Management and Economics, 19*(5), 493–502.

Gao, S., Low, S. P., & Nair, K. (2018). Design for manufacturing and assembly (DfMA): A preliminary study of factors influencing its adoption in Singapore. *Architectural Engineering and Design Management, 14*(6), 440–456.

Gao, S., Jin, R., & Lu, W. (2020). Design for manufacture and assembly in construction: A review. *Building Research & Information, 48*(5), 538–550.

Gerth, R., Boqvist, A., Bjelkemyr, M., & Lindberg, B. (2013). Design for construction: Utilizing production experiences in development. *Construction Management and Economics, 31*(2), 135–150.

Guedes, M. C., Pinheiro, M., & Alves, L. M. (2009). Sustainable architecture and urban design in Portugal: An overview. *Renewable Energy, 34*(9), 1999–2006.

Hill, R. C., & Bowen, P. A. (1997). Sustainable construction: Principles and a framework for attainment. *Construction Management & Economics, 15*(3), 223–239.

Jaganathan, S., Nesan, L. J., Ibrahim, R., & Mohammad, A. H. (2013). Integrated design approach for improving architectural forms in industrialized building systems. *Frontiers of Architectural Research, 2*(4), 377–386.

Kaner, I., Sacks, R., Kassian, W., & Quitt, T. (2008). Case studies of BIM adoption for precast concrete design by mid-sized structural engineering firms. *Journal of Information Technology in Construction (ITcon), 13*(21), 303–323.

Kendall, S. (1999). Open building: An approach to sustainable architecture. *Journal of Urban Technology, 6*(3), 1–16.

Kim, M. K., McGovern, S., Belsky, M., Middleton, C., & Brilakis, I. (2016). A suitability analysis of precast components for standardized bridge construction in the United Kingdom. *Procedia Engineering, 164*, 188–195.

KPMG. (2016). *Smart construction: How offsite manufacturing can transform our industry.*

Kuo, T. C., Huang, S. H., & Zhang, H. C. (2001). Design for manufacture and design for 'X': Concepts, applications, and perspectives. *Computers & Industrial Engineering, 41*(3), 241–260.

Kurokawa, K. (1977). *Metabolism in architecture*. Studio Vista.

Kurokawa, K. (2005). *Kisho Kurokawa: metabolism and symbiosis.* Jovis.

Laovisutthichai, V., Lu, W. S., & Bao, Z. K. (2020). Design for construction waste minimization: Guidelines and practice. *Architectural Engineering and Design Management.* https://doi.org/1 0.1080/17452007.2020.1862043

Lu, W., Tan, T., Xu, J., Wang, J., Chen, K., Gao, S., & Xue, F. (2020). Design for manufacture and assembly (DfMA) in construction: The old and the new. *Architectural Engineering and Design Management*, 1–15.

Maas, G., & van Eekelen, B. (2004). The Bollard—the lessons learned from an unusual example of offsite construction. *Automation in Construction, 13*(1), 37–51.

Meiling, J., & Sandberg, M. (2009). Towards a feedback model for offsite construction. In *Annual ARCOM conference: 07/09/2009-09/09/2009* (Vol. 1, pp. 291–300). Association of Researchers in Construction Management.

Owen, C., & Dovey, K. (2008). Fields of sustainable architecture. *The Journal of Architecture, 13*(1), 9–21.

Pasquire, C. L., & Connolly, G. E. (2003). Design for manufacture and assembly. In *11th annual conference of the International Group for Lean Construction* (pp. 184–194).

Peterseim, J. H., White, S., & Hellwig, U. (2016). Novel solar tower structure to lower plant cost and construction risk. In *AIP conference proceedings* (Vol. 1734, No. 1, pp. 070025). AIP Publishing LLC.

Pitt, M., Tucker, M., Riley, M., & Longden, J. (2009). Towards sustainable construction: Promotion and best practices. *Construction Innovation, 9*, 201–224.

Poon, C. S., Yu, A. T., & Jaillon, L. (2004). Reducing building waste at construction sites in Hong Kong. *Construction Management and Economics, 22*(5), 461–470.

Ramaji, I. J., Memari, A. M., & Messner, J. I. (2017). Product-oriented information delivery framework for multistory modular building projects. *Journal of Computing in Civil Engineering, 31*(4), 04017001.

Rausch, C., Edwards, C., & Haas, C. (2020). Benchmarking and improving dimensional quality on modular construction projects–a case study. *International Journal of Industrialized Construction, 1*(1), 2–21.

Royal Institute of British Architects (RIBA). (2019). *RIBA sustainable outcomes guide*. Retrieved from https://www.architecture.com/knowledge-and-resources/resources-landing-page/sustainable-outcomes-guide

Royal Institute of British Architects (RIBA). (2020). *RIBA plan of work 2020 overview*. Retrieved from https://bit.ly/2Rwz6RH

Russell, A. L. (2012). Modularity: An interdisciplinary history of an ordering concept. *Information & Culture, 47*(3), 257–287.

Sachs, J. D. (2012). From millennium development goals to sustainable development goals. *The Lancet, 379*(9832), 2206–2211.

Sassi, P. (2006). *Strategies for sustainable architecture*. Taylor & Francis.

Shi, L., Ye, K., Lu, W., & Hu, X. (2014). Improving the competence of construction management consultants to underpin sustainable construction in China. *Habitat International, 41*, 236–242.

Tam, V. W., Tam, C. M., Zeng, S., & Ng, W. C. (2007). Towards adoption of prefabrication in construction. *Building and Environment, 42*(10), 3642–3654.

Tan, T., Lu, W., Tan, G., Xue, F., Chen, K., Xu, J., Wang, J., & Gao, S. (2020). Construction-oriented design for manufacture and assembly guidelines. *Journal of Construction Engineering and Management, 146*(8), 04020085.

United Nations (UN). (2015). *Transforming our world: the 2030 Agenda for Sustainable Development*.

Wasim, M., Han, T. M., Huang, H., Madiyev, M., & Ngo, T. D. (2020). An approach for sustainable, cost-effective and optimized material design for the prefabricated non-structural components of residential buildings. *Journal of Building Engineering*, 101474.

Wong, K. D., & Fan, Q. (2013). Building information modelling (BIM) for sustainable building design. *Facilities, 31*, 138–157.

Xu, J., & Lu, W. (2019). *Design for construction waste management*. Paper presented at the sustainable buildings and structures: Building a sustainable tomorrow: Proceedings of the 2nd International Conference in Sustainable Buildings and structures (ICSBS 2019), October 25–27, 2019, Suzhou, China.

Index

© The Author(s), under exclusive license to Springer Nature Switzerland AG 2021
S. S. Y. Lau et al. (eds.), *Design and Technological Applications in Sustainable
Architecture*, Strategies for Sustainability, https://doi.org/10.1007/978-3-030-80034-5

Printed in the USA
CPSIA information can be obtained
at www.ICGtesting.com
LVHW022116270923
759263LV00005B/297

9 783030 800338